Swamp Rice Farming

Westview Replica Editions

The concept of Westview Replica Editions is a response to the continuing crisis in academic and informational publishing. Library budgets for books have been severely curtailed. Ever larger portions of general library budgets are being diverted from the purchase of books and used for data banks, computers, micromedia, and other methods of information retrieval. Interlibrary loan structures further reduce the edition sizes required to satisfy the needs of the scholarly community. Economic pressures on the university presses and the few private scholarly publishing companies have severely limited the capacity of the industry to properly serve the academic and research communities. As a result, many manuscripts dealing with important subjects, often representing the highest level of scholarship, are no longer economically viable publishing projects--or, if accepted for publication, are typically subject to lead times ranging from one to three years.

Westview Replica Editions are our practical solution to the problem. We accept a manuscript in camera-ready form, typed according to our specifications, and move it immediately into the production process. As always, the selection criteria include the importance of the subject, the work's contribution to scholarship, and its insight, originality of thought, and excellence of exposition. The responsibility for editing and proofreading lies with the author or sponsoring institution. We prepare chapter headings and display pages, file for copyright, and obtain Library of Congress Cataloging in Publication Data. A detailed manual contains simple instructions for preparing the final typescript, and our editorial staff is always available to answer questions.

The end result is a book printed on acid-free paper and bound in sturdy library-quality soft covers. We manufacture these books ourselves using equipment that does not require a lengthy make-ready process and that allows us to publish first editions of 300 to 600 copies and to reprint even smaller quantities as needed. Thus, we can produce Replica Editions quickly and can keep even very specialized books in print as long as there is a demand for them.

About the Book and Author

This first detailed ethnographic account of the Pahang Malay people of peninsular Malaysia focuses on the society's traditional agricultural system, particularly on its specialization in the production of rice on largely unmodified natural swampland. Dr. Lambert discusses the historical development of Pahang Malay rice farming, its dependence on indigenous knowledge of local ecology, and its adaptability to adverse conditions. Farmers experimenting with cultivars, adapting new technologies to local conditions, and using their own seed selection skills have over several decades substantially improved their rice yields. Dr. Lambert suggests that well-adapted indigenous farming systems found throughout the world should be studied and the adoption of these successful agricultural practices should be encouraged by governments and development planners.

Donald H. Lambert is a teaching associate in anthropology and political economy at the University of Texas at Dallas.

To the late
Tok Hitam binti Katid Gendut:
Orang Pesagi and friend

Swamp Rice Farming
The Indigenous Pahang Malay Agricultural System

Donald H. Lambert

NEW YORK AND LONDON

First published in paperback 2024

First published 1985 by Westview Press, Inc.

Published 2019 by Routledge
605 Third Avenue, New York, NY 10158

and by Routledge
4 Park Square, Milton Park, Abingdon, Oxon OX14 4RN

Routledge is an imprint of the Taylor & Francis Group, an informa business

Library of Congress Cataloging in Publication Data
Lambert, Donald H.
Swamp rice farming.
(A Westview replica edition)
Bibliography: p.
1. Rice--Malaysia--Pesagi (Pahang) 2. Agriculture, Primitive--Malaysia--Pesagi (Pahang) 3. Ethnology--Malaysia--Pesagi (Pahang) 4. Malays (Asian people)--Malaysia--Pesagi (Pahang)-- Economic conditions. I. Title.
SB191.R5L34 1985 633.1'8'095951 84-15326

Publisher's Note
The publisher has gone to great lengths to ensure the quality of this reprint but points out that some imperfections in the original copies may be apparent.

ISBN: 978-0-367-30479-9 (pbk)
ISBN: 978-0-367-28933-1 (hbk)
ISBN: 978-0-429-30808-6 (ebk)

DOI: 10.1201/9780429308086

Contents

Figures

Preface

The Pahang River basin of West Malaysia has contained a sizable population of Malay farming peoples for at least the last several hundred years. Many continue to live in villages along stretches of the river, and practice a mixed economy based on swamp rice farms, river and swamp fishing, cash from rubber and fruit production, with minor attention given to cattle and poultry rearing, kitchen gardens, and exploitation of natural forest products. While the Pahang Malay in recent decades are experiencing an incredible range of changes, mostly caused by Malaysian Government development programs in Pahang, and many are turning to newly available economic opportunities, the majority of them see their future in maintaining and adapting existing swamp rice and other indigenous resource utilization methods. While my study in the central Pahang village of Pesagi shows that most households are primarily supported by the production of rubber for cash, my informants took far greater pride in characterizing themselves as rice growers. In an ecological sense the highly adapted technique for producing rice and fish in the largely unmodified natural swamps bordering the Pahang river and coast is the Pahang Malay's niche specialization.

This book on the swamp rice agricultural system of the Pahang Malay follows closely my doctoral dissertation, and a report prepared on this subject for the Malaysian Government. Fieldwork in Pahang was carried out with support from a Fulbright-Hays Graduate Research Fellowship from January 1976 until March 1977. This current product is the result of continuing research since that time utilizing the data analysis facilities of the Quantitative Anthropology Laboratory and libaries at UC Berkeley, and revisions made while teaching anthropology at the University of Texas at Dallas.

While in Malaysia I received assistance from many people. I am obligated to members of the National

Unity Board and the General Planning Unit of the Prime Minister's Department, to the staff of the National Museum and the University of Malaya. My local sponsor was the Malaysian-American Commission on Educational Exchange (MACEE), which helped me solve countless problems and made my stay productive and enjoyable. A great deal is owed to the Daerah and Mukim officials for their support and assistance throughout the fieldwork period. I must especially thank Penolong Pegawai Daerah Encik Mohammad Ibrahim bin Abu Bakar, Wakil Raya'at Yang Berhormat Puan Sariah binti Kamiso PPM, Pengulu Ishak bin Haji Awang Lembek, Pengulu Wan Mohammad Ali bin Wan Haji Abdul Malik, and Ketua Kampong Mahmud bin Mat Taib PJK.

I am grateful for advice and encouragement from colleagues, especially William H. Geoghegan and James N. Anderson. Thanks are owed for help in preparing this manuscript to M.J. Tyler, Loel Miller, Linda Viloria, Dorothy Luttrell, Evelyn Stutts, Joan Seagraves, and the other staff at the Language Behavior Research Laboratory/Quantitative Anthropology Laboratory at Berkeley, and the School of Social Sciences at UT Dallas.

Of course this book would not have been possible without the interest and cooperation of the villagers at Pesagi. My family and I were treated with such warmth and friendliness that we feel a lasting emotional attachment to our many friends there. Everytime I work with the data, use the materials in my teaching or writing, or think of the fieldwork experience I long to see my friends and Pesagi again.

Donald H. Lambert
The University of Texas at Dallas
November 1984

1 Introduction: Indigenous Agriculture—Survival in a Modern Age

The Pahang Malay live along the coast and major rivers of the state of Pahang in West Malaysia. While a few Pahang Malay live in urban settings such as Temerloh or Kuantan, more than nine out of ten live in rural villages. This book is based on fourteen months of field work between January, 1976 and March, 1977 in the rural village known as Pesagi. In Pesagi each individual household practices a broad range of interrelated economic activities which villagers call kerja kampong (lit. 'village work'). This production complex, referred to as the indigenous Pahang Malay agricultural system, is comprised of cash and subsistence cropping, gathering of wild produce, hunting and fishing, various cottage industries, petty trading, cattle rearing, and maintaining small flocks of poultry. Before embarking on a detailed description of the studied village, I should like here to emphasize the importance of research on indigenous production systems, and highlight the unique benefits derived from using methodological and theoretical approaches in fieldwork which discover and record indigenous systems of knowledge.

One characteristic of the indigenous agricultural system of the Pahang Malay, is that over long periods of time it has consistently met the basic production needs of the local population. For example, while yields of rice, the major food crop, have varied greatly from year to year, farmers have always been able to offset this variation by carefully managing a diverse range of other productive activities such as fishing, home gardening and poultry rearing.

The lack of dependence on a single crop together with maintaining complex resource exchange networks and use of a broad range of microenvironments affords the Pahang Malay the capability for rapid adjustment to relatively small ecological and economic variations, and provides an inherent capacity to withstand great social and economic upheavals. Pahang agriculture indeed appears to be inherently adapted to withstand high frequencies of adverse conditions. Devastation and disruptive warfare throughout the 19th century and recurring with Japanese occupation and the "Communist Emergency" of the 1940s and 50s, traumatic flood and droughts about every 40 years with lesser episodes at least once per decade, changes in government, and unpredictable market prices are among the frequent disruptive episodes common to the Pahang scene. Nonetheless, the Pahang Malay have always maintained a secure and reliable level of production for local consumption. Never in recorded history has Pahang experienced famine, or even severe food shortages, in spite of frequent and sometimes extraordinarily disruptive calamities.

While in recent decades much effort has been expended to eradicate what were perceived as "backward" or "wasteful" traditional technologies, we now see that objective observers of some indigenous systems are so impressed as to argue that "indigenous knowledge should be taken seriously," (Belshaw 1980), and that "traditional systems" are ecologically functional and inherently rational (Igbozurike 1971). James N. Anderson (1979) at a recent International Symposium of Tropical Ecology held in Kuala Lumpur, concludes that indigenous agricultural systems such as the "Southeast Asian traditional home garden", have a very long history of providing people self-reliance and participation in their own development, and that there is little doubt that there are considerable potential benefits from making improvements to old, long-evolved, and time tested indigenous systems.

PROBLEMS OF MODERN AGRICULTURE IN PAHANG

While agricultural specialists have generally assumed that "modern methods" are of outstanding merit on most counts, cases are certainly not absent in which indigenous agricultural methods have proven superior to modern ones. The 'swamp rice' (or _padi paya_) system of Pahang, which this report describes in detail, is just one example of a superlative indigenous adaptation to both short-term and long-term environmental fluxations. Repeated attempts to establish modern technology on central Pahang swamplands, especially those in the Temerloh District, have not been encouraging. Many dams, drainage ditches, and other water control works built by government agencies and foreign to the Pahang system have failed because of the heavy rates of siltation or erosion from floods. Agriculture Department literature from the thirties shows that test station experiments on rice varieties conducted at Pulau Tawar,

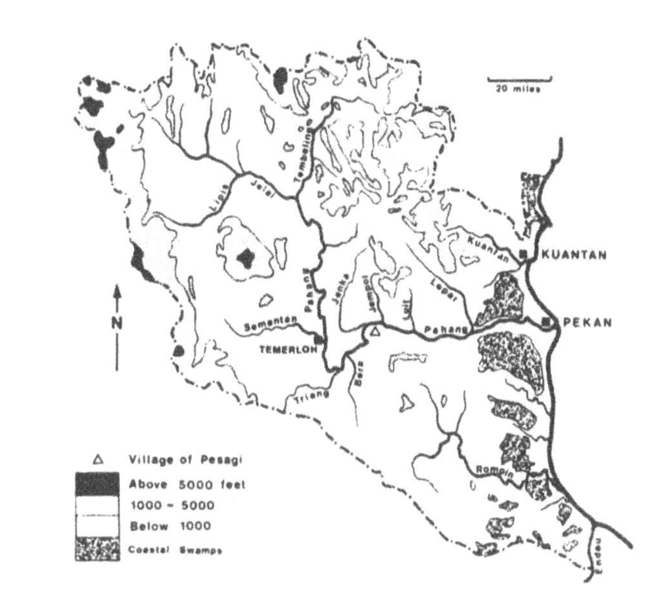

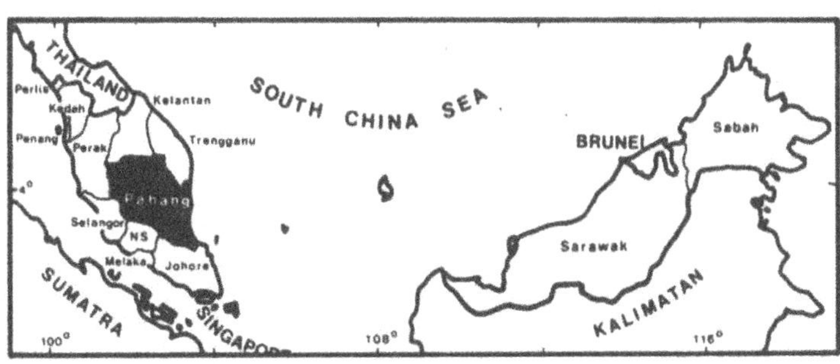

FIGURE 1 MAP: PAHANG AND MALAYSIA

Kerdau, Mengkarak, and Chenor were not successful in finding "improved" rice varieties suitable for swamp conditions. In most cases, introduced test varieties performed with great inconsistency or produced lower yields than traditional varieties (Birkenshaw 1941; Jogoe 1939). In the riverine areas of central Pahang technicians found "modern" rice production methods quite unsuited to the extremes of the local environment, and eventually gave up on attempts to improve indigenous rice cropping methods. (A. O. Report 1929-37).

In another example, that of rubber (<u>Hevia brasiliensis</u>), a crop introduced to Malaya by British corporations using estate and plantation technology, farmers using local methods have for many decades had a significant competitive advantage under economic and ecological crisis conditions. In Ooi Jin-Bee's comparison of smallholders' rubber-growing strategies to those used on plantations, he argues that smallholders are more able to avoid risks and undesirable costs than plantation owning corporations. He writes that

> practical experience in Malaya and other tropical countries has shown that a well-organized system of peasant farming, because of its greater flexibility is better able <u>to withstand crisis conditions</u> than an economy based on plantation agriculture. For example, during the Great Depression of the 1930s, the rubber plantations were badly hit while the peasant rubber smallholders simply left off tapping and turned to other alternative occupations without having to worry about heavy overhead costs. Again, an economy based on peasant farming gives greater economic and social stability to the country during a major depression because farmers can always turn from growing cash crops to growing foodcrops. No mass unemployment need follow such a depression, unlike the position in a country which is mainly dependent on plantation agriculture [emphasis mine] (Ooi 1963:195).

Unfortunately history has had little impact on the direction of current government managed rural development projects, as evidenced by the fact that most rubber, oil palm and other modern projects are largely systems which follow the plantation design. A more practical approach would be to model development projects on selected time-tested and well-adapted indigenous forms.

Other problems with introduced modern agriculture are highly visable and regularly reported in Malaysian newspapers and elsewhere, such as widespread pollution of the environment by agricultural chemical affluents

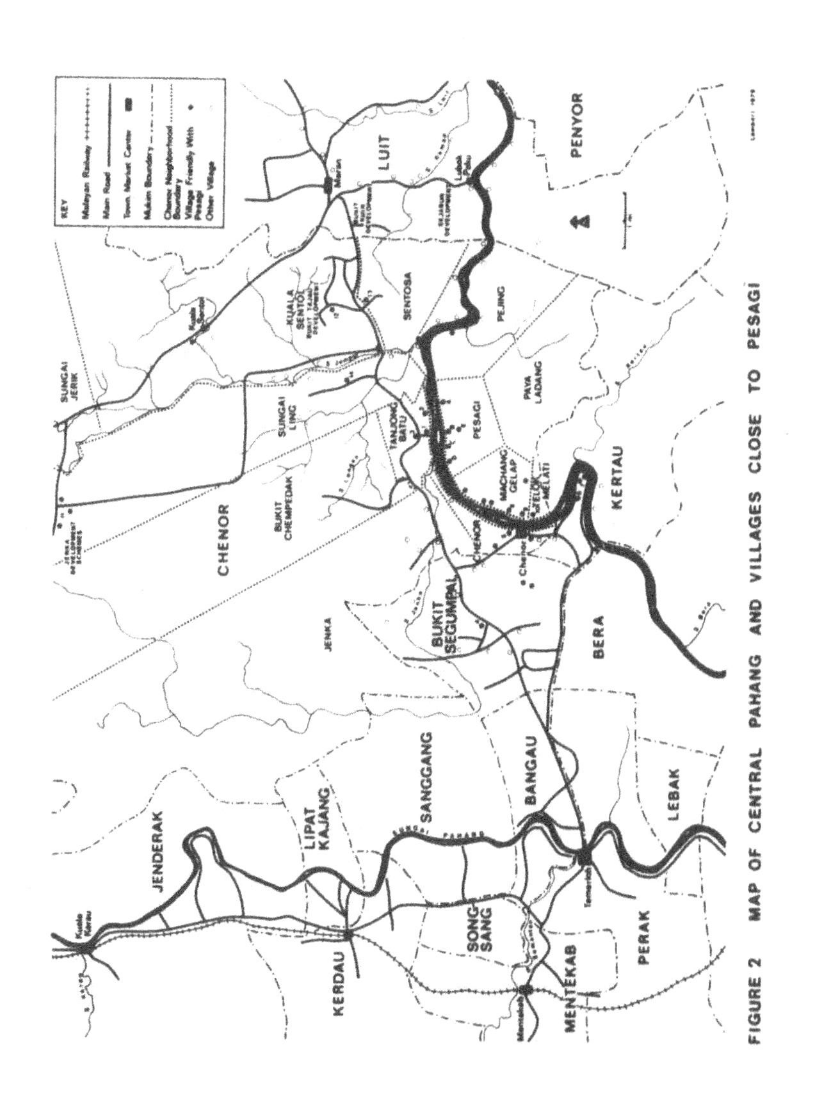

FIGURE 2 MAP OF CENTRAL PAHANG AND VILLAGES CLOSE TO PESAGI

of the environment by agricultural chemical affluents from processing factories (NST October 2, 1976; (NST October 31, 1976; NST November 1, 1976; R.N. 1975; Ho 1976a, 1976c; Malaysian Government 1971:253); the inherent weaknesses of mono-crop and "green revolution" systems such as population displacement and low or unreliable family incomes (Asian Development Bank 1971; Ho 1980a; Igbozurike 1971; Peyman 1980a, 1980d, 1980d); and high levels of social and economic inequity (Das 1980a; NST August 24, 1976; Peyman 1980c). I plan to write a fuller account of these problems in a later paper.

Surprisingly, the advantages of indigenous production systems --including dependable production and conservation of natural resources --are largely unrecognized or not seen as valuable by professional development advisors, agriculturalists, and government civil servants. I submit that one means of resolving some of the development problems in Malaysia and elsewhere, is to enhance and modify the existing indigenous mixed-crop systems. These systems, as the Pahang material presented in this study will demonstrate, supply nearly all the farm family's needs: staple and supplementary fresh foods, condiments and medicines, fuels, building materials, and cash crops for purchasing other necessities and services. The major short-comings, -- low-level production and land shortage, inadequate marketing and transportation facilities, and lack of research, educational, and extension programs appropriate to small scale production systems and family managed businesses -- could be overcome if national priorities were shifted to utilize human, natural and capital resources efficiently and equitably amongst traditional farm families.

In an agroecological sense the traditional agriculture of Malay and other indigenous groups is more modern than recently introduced mono-crop schemes, because of greater flexibilty, diversity, integration, and stability (cf. Anderson 1979). There is much to be said for the observation of the Conference on Productivity and Innovation in Agriculture in the Underdeveloped Countries, that "greater benefit in relation to costs will be obtained in many cases from an increase in the productivity of already cultivated land [mostly indigenous] than by attempting to bring new lands under cultivation (Millikan and Hapgood 1967:40).

IMPORTANCE OF STUDYING INDIGENOUS SYSTEMS OF KNOWLEDGE

Throughout the material which follows I emphasize, when known to me, the perspective and knowledge of the people studied. This emphasis is a courtesy to the people concerned; accents human achievements, needs, and resources; aids in understanding the ongoing natural development process as well as providing crucial and elementary information for adapting technology to new settings; records and preserves valuable local knowledge; leads local persons to an increased awareness and pride in their own achievements; and provides a basis from which the studied people can be better appreciated by other members of their own and other societies (cf. Brokenshaw _et al._ 1980:8).

Hopefully this study of the Pahang Malay agricultural systems is a step towards finding ways to increase levels of production without overly sacrificing major adaptive characteristics, and will concurrently make possible a minimal disruption of the functional and meaningful social-cultural heritage of these people within their own given ecological setting. It should lend itself to the design of rural development projects which increase social and economic equity, and involve people in expanding opportunities to make meaningful decisions about the direction of their lives. This description of Pahang Malay agriculture should promote an appreciation of the valuable resource embodied in existent rural folk themselves, as well as appreciation of rural folk-knowledge. Certainly this is a worthwhile task, and an invitation for other researchers to study this time-tested reliable system which has made possible the permanent and successful settlement of the river valleys and coastal plains of Pahang for the last several hundred years. Such an enquiry is timely and relevant in a Malaysia where the national economy, not to mention national security, is predominately dependent on the knowledge, skills, labor and products contributed by farmers.

2
Pahang and Pesagi History

This chapter will explore the history of the Pahang Malay, and the village of Pesagi, with emphasis on the development of present day economic behavior.

Prehistoric sites show that Pahang was populated from very early times. M. W. F. Tweedie, a major investigator of Pahang prehistory before the 1960's, concluded that a thinly scattered neolithic culture was present in the interior as early as the beginning of the Christian era, which practiced shifting cultivation, possibly grew yams (especially Dioscora alata) and rice, and used tools suggesting the capability to build timber houses and dugout boats (1953). More recent work by Frederick L. Dunn (1964, 1966, 1970) has uncovered a cord-marked pottery stratum radio carbon dated at 4,800 ± 800 years B.P., directly succeeded by a "Late Neolithic" stratum. In some stratified sites this cord-marked pottery is proceeded by a considerably earlier "Hoabinhian-like" stratum, containing assemblages of flaked-core tools accompanied by utilized flakes, grinding and pounding tools (Dunn, 1970; c.f. Peacock and Dunn 1968). These data suggest that Pahang is to be included in the general Southeast Asian Late Pleistocene and Holocene cultures which Wilhelm G. Solheim II (1968), Chet F. Gorman (1969) and others (c.f., Dunn 1970) have hypothesized as extending throughout much of South China, parts of East Asia, mainland and insular Southeast Asia, and into the Pacific Islands. The most extensively researched Hoabinhian culture sites are Spirit Cave and Non Nok Tha in Thailand, where evidence suggests the domestication of plants as early as 15,000 B.C., pottery from 10,000 B.C., domesticated beans from 9,000 B.C., dugout canoes common by 5,000 B.C. with the outrigger and hence sea voyaging by about 4,000 B.C., and rice cultivation from 4,500 B.C. (Solheim 1971). Given that the amount of fieldwork in Pahang is as yet far from adequate, it is not possible to estimate when Pahang's

Hoabinhian period began, whether the influence was continuous over time, the various migrations or identities of peoples involved, or the genetic, technological or other linkages between these early inhabitants of Pahang and the Pahang Malay of today.

Unfortunately even Pahang's recent past, particularly before the fourteenth century, is virtually unknown as while there is voluminous early literature, it is extremely obscure. The earliest documents thought by scholars to contain references to the Malay Peninsula include Indian texts from the latter part of the first millenium B.C., Chinese texts from about 4 A.D., Arab texts from 8 A.D., and Latin texts from 43 A.D. (Wheatley 1961). Most of these early documents are problematical because of their second-hand nature. The authors were not travelers, but compilers. Thus, they contain much which current scholars believe to be rumor, fantasy, or repetition of previous sources. Also, their utility is limited because modern scholars are unable to accurately determine the precise locations to which place names refer. Wheatley tells us, that "...so wide is the gulf between the geography of ancient and modern Malaya that there has been only a tenuous continuity of nomenclature from early times to the present" (Wheatley 1961:vii).

Only from the fourteenth century on do we have reliable sources of information. At that time, merchants, emissaries, and scholars began to write of their Malayan adventures, and a few indigenous written sources began to appear (e.g. Sejara Melayu and court records in the mid-16th century). About 20 years after Columbus sailed to the new world, European nationals began establishing colonies and writing about the Far East. Due to the low reliability of the earlier materials, I will concentrate on the last two or three hundred years, which is also the period most relevant for understanding the present culture.

THE PAHANG PEOPLE

The Orang Pahang ('Pahang People') of today define themselves as the descendants of people who have always lived within the area presently known as the State of Pahang. Throughout this report the term "Pahang People" is used to refer to Malay and Orang Asli[1] who 'are born in and have ancestors from Pahang' (orang asal di Pahang), and specifically for persons possessing several generations of genealogical and historical ties to Pahang. The term "Pahang Malay" is always used to refer only to that portion of Pahang People who are Malay, never for the many Malay from other states who have recently moved to Pahang. The

Pahang Malay are Muslim, speak the Malay language, and the majority live in riverine or coastal areas. The Asli tend to live away from main rivers except the Endau and Rompin, and in most cases are not Muslim.

The major increase in the population of Pahang during the twentieth century (see Figure 3) is due to in migration. In the 1880s the population was about 60,000 and made up almost entirely of Pahang People, that is, Pahang Malay and Asli. By the late 1930s, Chinese, Indians, and immigrant Malay already made up about one half of the total population. Pahang People refer to these immigrants as _orang luar_ ('aliens'). Since World War Two the pace of immigration has increased so rapidly that by 1975 outsiders were twice as numerous as Pahang People. The present day population, projected from these figures is estimated at more than 600,000, about 60 percent of which is Malay (one third of these are recent outsiders), 30 percent Chinese, and 10 percent from other ethnic groups (including Indians, Pakistanis, Thai, and Europeans). In terms of distribution, rural Pahang is predominately Malay, with the other groups concentrated within or close to the major towns of Kuantan, Pekan, Maran, Temerloh, Mentakab, Raub, Bentong, Karak, Triang, and Jerantut.

The Pahang Malay State

There is evidence of small chiefdoms existing in the Malay peninsula for at least 400 years before European arrival (Lam 1964; Purcell 1965; Wheatley 1961). One of these chiefdoms, administered from a port at the Pahang River mouth, was first mentioned in Chinese documents dating from the 12th and 14th centuries (e.g. cited in Wheatley 1961). By 1414, Pahang had obtained considerable economic importance, and was recorded as exchanging tribute with China, and described by the emissary Cheng-Ho as follows:

> The ground of this country [Pahang] is constantly warm; rice grows abundantly. They boil sea water in order to make salt and brew the sap of cocoanuts to get wine. Superiors and inferiors are very intimate together, and they are neither thieves or robbers. But they are deluded by spirits and ghosts of which they carve images of perfumed wood, to which they offer human victims in order to avert calamities and to pray for blessings (Schlegel 1899:42).

The Chinese accounts record extensive lists of trade items, and suggest that Pahang was a well organized commercial state. The major exports were camphor, several kinds of incense wood, laka wood, and a perfume known as tai-pai. From China Pahang regularly imported iron and copper cauldrons, cotton and silk cloth, and lacquer-ware (Wheatley 1961).

Of course it required an elaborate social and economic network extending well inland from the coast in order to effectively procure and transport these export commodities. All the exported products occur only in forest, and some, such as camphor (Dryobalanops aromatica), grow best at elevations found far from the coast (cf. Burkhill 1966:876).

From the fifteenth to seventeenth centuries numerous Malayo-Muslim sultanates and chiefdoms were organized in that area of the archipelago along the coasts of Sumatra, Borneo, the Malay peninsula, and some adjacent islands (Steinberg 1971:73-79). On the peninsula these included the Malay states of Perak, Kedah, Pahang, Johore, and Melaka. Elsewhere, Acheh was located at the northern tip of Sumatra, there were several small Minangkabau states in western Sumatra, and in the eastern portion of the Indonesian Archipelago there was Brunei in north-west Borneo, Sulu in the southern Philippines, and the Bugis state of Makassar in Suluwesi (i.e. Celebes).

Gradually, beginning in the early sixteenth century, European nations acquired control of Southeast Asian trade, and over a period of 350 years weakened the economic and political foundations of Malay and other indigenous states of the region (Purcell 1965; Bassett 1964; Gullick 1963; Hall 1968; Steinberg 1971). For several decades before European arrival, the Malay State of Melaka had assumed the dominant economic position in the western archpelago (Meilink-Roelofsz 1962; Anderson and Vorster 1976; Bremmer 1927; Mills 1930). The Europeans were primarily interested in monopolizing the spice trade with the Arabs and India, and gaining control over the trade routes to China. Melaka was overthrown by the Portuguese in 1511, taken by the Dutch in 1641, and became a British port city in 1786.

Pre-colonial Pahang of the seventeenth to the nineteenth centuries comprised four territories: Pekan, Chenor, Temerloh, and Jelai (Linehan 1936). Each was ruled by a chief (orang besar, lit. 'big man') and a sultan resided in the Pahang River port at Pekan. During much of the Pahang state's history the nobility and territorial chiefs were divided by struggles for internal trade and power, and their followers were frequently involved in feutalistic battle deployment, attacking rival territories, killing or driving people from their holdings, and destroying crops (Linehan 1936:45-55). In the absence of statewide popular support, the sultan could not control some major

FIGURE 3 PAHANG POPULATION BY ETHNIC GROUP

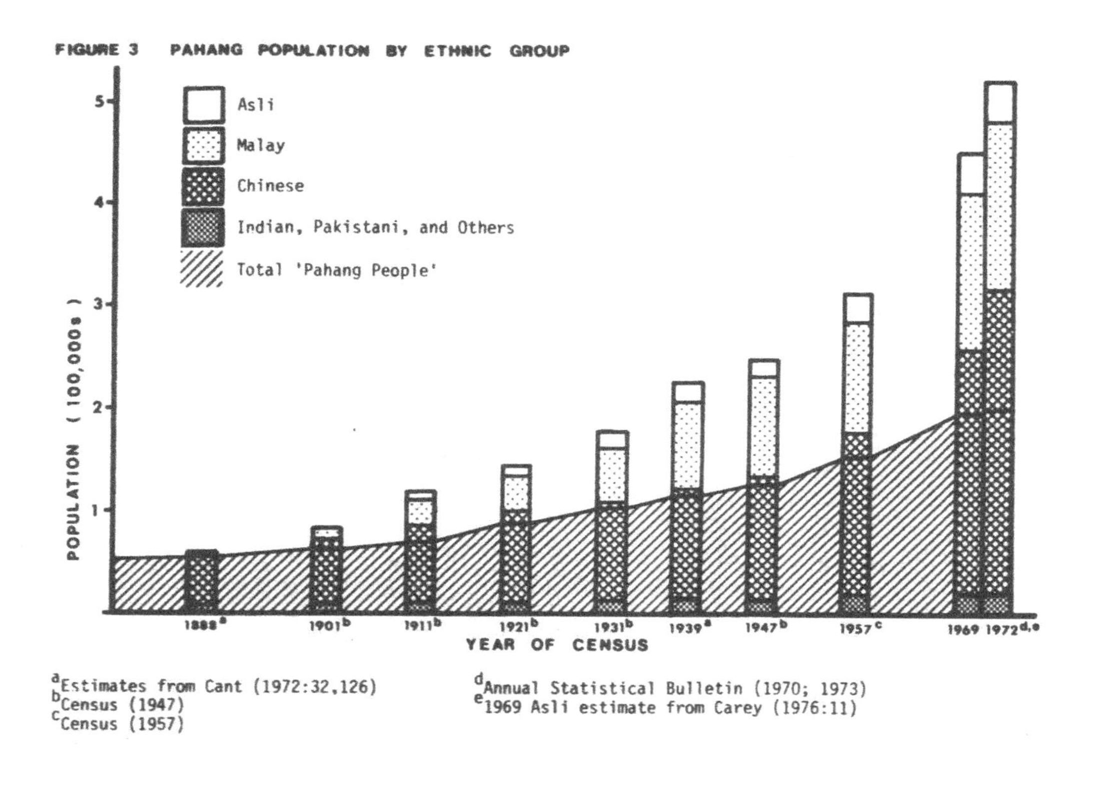

[a]Estimates from Cant (1972:32,126)
[b]Census (1947)
[c]Census (1957)
[d]Annual Statistical Bulletin (1970; 1973)
[e]1969 Asli estimate from Carey (1976:11)

international trade goods such as gold, which was mined far from Pekan in the Jelai and Temerloh Territories. Further evidence of Pahang's lack of unity is that chiefs in the interior areas maintained trade routes that by-passed the Pahang River and its port at Pekan. These routes (see Figure 4) consisted of passes through the hills from Pahang rivers to the headwaters of rivers in surrounding Malay states. The interior chiefs commanded wealth and large followings of supporters both inside and outside of Pahang, which at times completely overshadowed the power of the sultan and made him de facto ruler of only a small coastal region.

In the first half of the nineteenth century Pahang was ruled by Bendahara Tun Ali, whose undisputed jurisdition extended to all four territories, with a total Malay population of about 51,000 (Linehan 1926:138). This accomplishment was in part due to the fact that the interior Pahang chiefs were no longer able to maintain their powerful trading positions after the surrounding Malay states of Perak, Selangor, Negri Sembilan, Melaka, and Johore fell under British control.[2]

As the interior territories were forced to give up their trade routes and trade partners in neighboring states, and to trade down the Pahang River instead, the state and the Pahang Sultanate grew in unity but not necessarily in strength. Much of the increased Pekan trade had already fallen under control of the newly established British trading center at Singapore: the major market for gold, tin, and exotic products of the interior peninsula forests including benzoin, resin, and rattan.

The interpretation of historians and others differ concerning the nature of the relationship between local villagers and the state-wide trade network of the last century. Linehan (1936) says that for utilitarian purposes such as mining, warfare, domestic service, forest product extraction, and agricultural production for personal use, the chiefs and noble classes conscripted the labor of their subjects. He feels that even though Pahang in general experienced increased stability and prosperity, this benefited the ruling class at the expense of villagers. Cant (1964) supports this interpretation, writing that the unemployed nobilities' stranglehold on the use of their subjects as laborers served to thwart agricultural development and recovery from earlier disruptions. Both of these historians were heavily influenced by the eyewitness account of Pekan by Munsi Abdullah (1893:14-28) in 1838, who records that rice was being imported, that foods of all kinds were scarce, and that the population lived in great poverty. Internal feuding, excessive

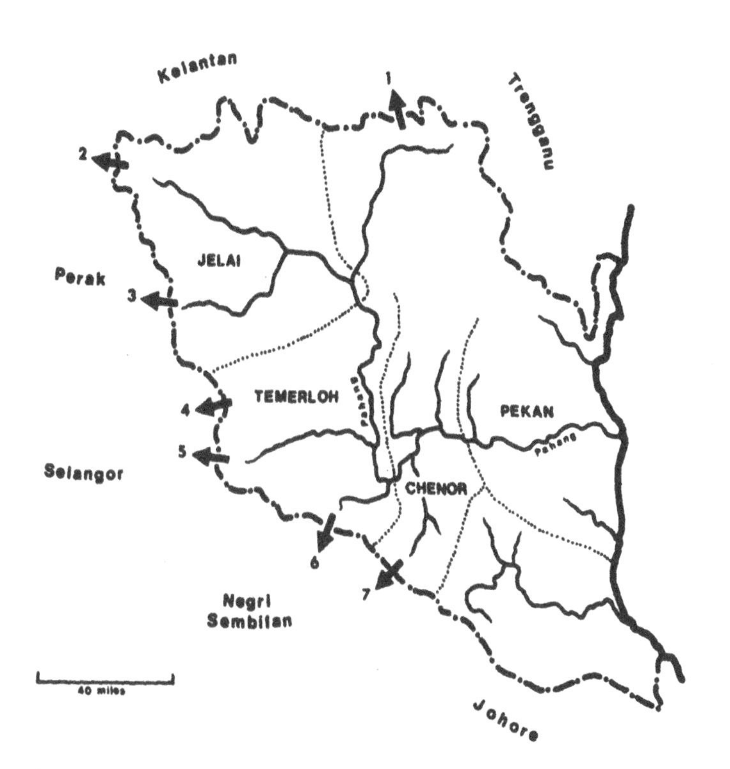

Trade Routes:

1. Ulu Tembeling to Ulu Kelantan
2. Ulu Jelai to Kinta Valley, Perak
3. Ulu Lipis to Ulu Selim, Perak
4. Ulu Teras to Kuala Kubu, Selangor (Semangkok Gap)
5. Ulu Semantan to Ulu Gombak, Selangor
6. Ulu Triang to Ulu Muar, Negri Sembilan
7. Ulu Serting to Ulu Muar, Negri Sembilan

(Sources: Cant 1972; Linehan 1936)

FIGURE 4 BOUNDARIES AND TRADE ROUTES OF THE FOUR TERRITORIES IN THE 19TH CENTURY

taxation by both ruler and chiefs, heavy import and export duties, debt-slavery, and corvee labor, were believed by Abdullah to have been burdensome.

However, the report of Abdullah is contradictory. While he stressed the level of poverty on the one hand, an impression made vivid by his account of "messy", "smelly", and "mosquito infested villages", he failed to recognize that these in themselves are not indicators of poverty, but are in fact natural and normal conditions of the coastal swamp environment. Moreover, Abdullah contradicts his own report of great poverty by noting that people were well dressed in cloth imported from India and Europe. Abdullah also describes a thriving upriver agricultural economy supplying Pekan and coastal regions with coconuts, areca nuts, yams, taro, sweet potatoes, sugarcane, and bananas.

Some scholars argue that Malay rulers did not excessively exploit their subjects, but that both ruler and subject operated within a system marked by reciprocity. Anderson and Vorster (1976:4,44) see the relationship between villager and ruler as mutually beneficial, since political power during this and earlier periods was in fact largely predicated on the economic bonds between ruler and subject. They assert that local villagers would logically participate in the primary collection and transportation of trade goods to export centers on behalf of their rulers, because such cooperation brought increased economic and political opportunities. Pahang villages were dependent on external goods such as cloth, metal goods, and salt, so that they were to a greater or lesser degree linked to the sultan's agents and local chiefs.

We can conclude from this, that, in nineteenth century Pahang economic effort rather than being primarily divested towards the production of export goods for the sultan was in fact quite diverse, with extensive local subsistance production, some well developed commercial agriculture at least near Pekan, and a state-wide system of marketing and bartering local crop surpluses and forest products for imported goods. This entire state-wide economic system was mediated by the sultan, the territorial chiefs, and numerous district and village level agents.

In 1857, the death of Bendahara Ali marks the beginning of the decline of the Malay state of Pahang. The Bendahara's sons Mutahir and Ahmad both claimed the right of succession, and led the territories in a disastrous civil war lasting six years. While Ahmad eventually won, was proclaimed Sultan, and ruled until his death in 1914, the state remained impoverished and disorganized, and never recovered economically or politically. In fact, by this time, reorganizing Pahang under a Malay ruler was impossible, since virtually all of the international trade on which the state leadership depended had to pass through British controlled Singapore and Melaka. By the mid-nineteenth century, the British East India Company was the paramount power in the Malay Peninsula, and before the century ended, its monopolization of trade and power brought about "the extinction or subjugation of virtually all indigenous political authority throughout the Malay world" (Steinberg 1971:135).

THE BRITISH ADMINISTRATION

In 1826 the British East India Company established the Straits Settlements of Penang, Melaka, and Singapore as collection centers and ports for regional trade. Originally they had little intention of penetrating into the Malay Peninsula itself, but by 1874 rivalries among the Malay States were seen as serious threats to the Company's interests. The fear of possible trade disruption, risk to food supplies, increasing Siamese interests in the southern peninsula, and the problem of protecting the rights of growing numbers of persons living in Malay States but born in the Straits Settlements (therefore British subjects), were among the official reasons given for establishment of the Resident System (Kennedy 1962). A Deputy Undersecretary of State in the Colonial Office, Sir Charles Jeffries, claims the move was made to ensure that mineral resources did not fall under control of rival powers (Jeffries 1956:82). By 1895 British Advisers, called Residents, were appointed to oversee the local governments in Perak, Selangor, Negri Sembilan, and Pahang.

In the decades immediately before and after the British takeover of Pahang, it was common practice for Europeans to organize expeditions to assess the natural resources, social and political conditions, and locations of legal boundaries. These included the Russian ethnologist Baron M. de Mikluho-Maclay (Mikluho-Maclay 1878; Skinner 1878), and a British government survey party headed by D. D. Daly (1882), who were frustrated on finding that local chiefs would not lead them to the gold mines. However, using intrigue and dealing with rival chiefs, European mining companies not only discovered the location of gold fields, which were thoroughly explored by 1885 (see Cameron 1885 and Cant 1964), but also gained concessions which disregarded the traditional rights of local chiefs (Thio 1957).

With the establishment of European businesses in Pahang, the British felt their own self-interests could only be protected by a Pahang Residency. The first attempt at negotiating a treaty in 1885 was turned down by Sultan Ahmad, but the British Representative Sir Frank Swettenham did manage to visit much of Pahang to take reconnaissance of conditions and rally support from the local population (Swettenham 1885). In 1887 the treaty attempt was successfully repeated by Hugh Clifford, who spent several months traveling, and published the first book length report of life in Pahang (Clifford 1903). These and other descriptions of Pahang (e.g. government botanist H.N. Ridley 1894; Brown 1913; and Daly 1882) described a society with small-scale village economies of rice, fruit, and poultry, with some limited rearing of cattle, washing gold, and manufacture of mats for local trade. Each also described the richness of the soils, the promise of tin and gold for mining, and the ideal climate for agricultural production. One major result of this early exploration and description of Pahang through European eyes, is the entrenchment of a grandiose image of Pahang which continues until this day -- in the popular mind Pahang is perceived as being virtually unpopulated, its people as being politically weak and culturally backward, and its rich resources as being almost incomprehensively extensive and available for the taking.

In 1888 J. P. Rodger became British Resident, the legal adviser to the Sultan of Pahang, with control over the entire traditional political hierarchy, in all administrative matters except for Malay religion and adat ('customary law'). The British Administration divided Pahang into five districts, each placed under control of a District Officer whose role was that of tax collector and magistrate. European officials were placed in charge of administrative departments such as Finance, Lands, Mines, Police, Prisons, and Education (Belfield 1902:4). The loan of Sikh police from Perak and Selangor helped to insure order. Increasingly, power and jurisdiction were prescribed by the new administration, which shifted responsibility from the Sultan to the Resident and his District Officers.

Pahang's internal politics went through stormy times during the early days of British administration. The Malay elite in particular, were dissatisfied with their reduced rights and privileges brought about as the British increasingly suppressed slavery, regulated the use of corvee labor, placed the Sultan and his chiefs on fixed allowances, withdrew the chief's customary rights of taxation, and regulated land tenure (Linehan 1936:129-38).

From 1891 to 1894, some interior chiefs gathered large followings and opposed the new government. An account of fighting that occurred near Pesagi, narrated to me by Mat Keteh bin Imam Taib from the Jempul River, describes battles that pitted the 'People of Pekan' (i.e. The British Force which included British and Sikh soldiers, the Sultan of Pahang, the Orang Kaya of Chenor and Temerloh, and the District Officer of Temerloh) against the 'People of Semantan' (i.e. The Anti-British Force which was led by Semantan Chief Dato Bahaman, and the warrior Mat Kilau). The people of Mat Keteh's village fled before the invading forces to an inland sanctuary up the Jenka River (whose mouth is at Pesagi), and from there inland to the forests of Bukit Segumpal. This same battle is described by the historian Linehan (1937:152-3) as the one in which the Orang Kaya of Chenor was killed in June 1892. Several months later the British sent a retaliatory Malay force up the Jempul River and succeeded in killing a minor rebel leader, the Panglima Muda; and my informant Mat Keteh recounts fleeing again with about 50 men, women, and children into the forest during heavy monsoon rains. Similar battles were fought elsewhere in Pahang until 1894, when the British forces drove the final rebel leaders into exile.

Invasion and warfare had long-term social and economic consequences for local villages. Typically, successful invasion of one Malay political unit by another resulted in killing the adult men, enslaving of women and children, destruction of houses and crops, cutting down of fruit trees, and killing of cattle. The ecological effects and related economic disruption in attacked villages lasted for many years. These war tactics served to preclude the possibility of a given geographical region mounting any retalitory opposition force for many years. This kind of military tactic also helped to define an adaptive pattern of survival and recovery characterized by reliance on short-term crops, low labor use, lack of dependance upon draft animals, a limited marketing system, and small networks made up of close family members.

The British administration established land registration in 1889 to provide security of title and ease of transfer, a process which brought radical changes to the local Malay economy. By 1895 the British began restricting occupation of new lands, and by 1900 the use of forest for ladang ('swidden farms') was prohibited. This placed detrimental limits on the Malay economy, which at that time was partly dependent on shifting cultivation. The new land policies brought local shortages of land, decreased the fallow period of the ladang system, and resulted in degradation into grasslands. The concurrent drop in agricultural productivity, following on the earlier economic and political collapse, had a further regressive and demoralizing effect that lasted until villagers successfully adapted to cash crops some two decades later.

Another great change occurred as transportation and communications shifted away from the use of rivers towards the use of road and railways. By 1921 the rail line had pushed into Pahang, linking Singapore with European controlled mines and rubber plantations in the north. Within a few years, roads connecting Temerloh and Kuala Lipis to Selangor in the west, and to Kuantan on the east coast, were also completed. These roads became the loci of rapid development, of primary benefit to Chinese, Malay, and Indians from outside Pahang (hence the influx of outsiders mentioned earlier and illustrated in Figure 3). Before roads were developed, the river dwellers participated in regional trade, to a greater or lesser extent, as they were integrally situated on the only trade routes through the state. During the early days of British rule it was remarked that central Pahang, along the river in the Temerloh District where wet rice was grown, was more advanced than any other area in the state (cf. Skinner 1884:54, Daly 1882:339). But within a short time, after the state was opened to outsiders, Pahang People began to suffer economic decline. Roads channeled the flow of resources, ideas, political power, trade, and modernization to the migrants who settled along them, rather than to the riverine Pahang People. Roads facilitated establishment of administrative and market towns, attractive to a population with more sophisticated commercial, industrial, and political interests, and composed largely of migrants from other parts of the Peninsula.

In 1926 the most destructive flood in Pahang history occurred (Winstedt 1927). Property loss was obviously greatest for river dwellers, and this served to magnify the already existing differences between them and the road settlers. Officials reported almost total destruction of buffalo, houses, and crops in some riverine areas. The annual reports of the Agricutural Office in Temerloh District noted a flood-related decrease in coconut and rice production which continued for many years (A. O. Report 1929, 1930, and 1935).

From about 1910 until the Second World War the Agricultural Office maintained extension programs in the river villages. There were many attempts to encourage the development of small scale cash crops, including coconut, fruits, coffee, tapioca, areca nut, bananas, various vegetables, and kapok. Following the 1926 flood, several villages in the Pesagi area, were encouraged by the Agriculture Office to plant kapok (over an area of 270 acres), with the incentive that European exporters were eager to purchase the crop (A. O. Report 1930). Unfortunately for the villagers, except for coconut (which remains a minor crop), these experiments failed for lack of markets, for lack of quality, and for low levels of production (A. O. Reports 1929-36). Other than distribute planting materials, the British government did little which lead to actual agricultural improvement in Malay villages. Indeed, by this time the Administrtion was preoccupied with estate production of the most rapidly expanding product of the Malayan colonial economy -- rubber -- and the Malay villages were largely neglected and expected to subsist by traditional means.

The Colonial Office wanted no competition in the production of major cash crops, advocating that the Malay continue a primarily subsistence economy with only a minimal participation in dominant cash crops. The Agricultural Officer of Temerloh District in 1913 published the following highly idealized view of the existing Malay way of life:

> Now that the Natives have been induced in their own interests to give more attention in the cultivation of the coconut...it would be really difficult to imagine anything more ideal for Malays than these Native holdings, which with the adjacent padi land appear to yield all they can possibly want that is needful for their welfare...there is always a convenient and large attap building on the premises for the owner and his family and what with coconuts, fruit trees, padi land, etc., the products from which alone should prove ample for the support of himself and his family, grazing ground for the buffalos and sheep, breeding of fowls, I am inclined to quote from one of my annual reports, that I am of the opinion, a Native, with comparatively small means who owns his 5 or 10 acres of land on his kampong property kept and cultivated is in his own way as well and comfortable off as the more wealthy owners of large estates (Brown 1913:227).

While the Administration gave no encouragement to rubber planting by the Malay, villagers who observed the rubber price boom of 1910, and the frantic rush of foreign companies to plant rubber during the 15 to 20 years following were quick to recognize advantages of the new cash crop. While large plantations suffered extreme difficulty from low and fluctuating prices which hindered expansion (see Appendix G), the Malay farmer's small-scale operation and reliance on a diversity of many different crops, made possible a steady increase in Malay smallholder rubber production. By 1940, the acreage of rubber planted by smallholders represented 40 percent of Malaya's total rubber acreage (McHale 1967: 73), and the percentage has steadily increased to 58 by 1977 (Gill 1977). McHale gives the following analysis for this expansion:

> The rapid increase in both acreage and output of Native smallholders in Southeast Asia stemmed from a variety of circumstances, including the following:
>
> (1) the rapid discovery of an "ideal" complementarity between the existing pattern of peasant swidden agricultural practices in hilly terrain and rubber planting,
> (2) the excellent prospects of an extremely attractive cash income on the basis of relatively small inputs of money, capital, or labour,
> (3) freely available rubber seeds after 1910,
> (4) an abundance of land in both Malaya and Sumatra, and
> (5) the rapid spread of basic rubber tapping and coagulation skills.
>
> Several other factors also played a part in the development. Once planted, rubber trees required little care to survive and could be abandoned (as many were in periods of low prices) for years at a time with no loss. Once they reached tapping age, rubber trees represented a latent capital asset, yet they did not represent tied up capital, involving no long-term labour or managerial commitments and merely required limited and easily learned skills to be transformed into actual productive assets. To the smallholder, a stand of rubber trees provided not only a means of realizing windfall incomes during periods of high prices, but also an immediate source of cash income on a literally daily basis all year around (McHale 1967, 66).

Between 1921 and 1939, the British government, fearful that Malay villagers would become landless dependents as the rubber industry expanded, began restricting estates from buying up Malay village lands. This policy also made it difficult for the Malay to acquire land anywhere outside their own reserve area. The area set aside, as Malay Reserve Lands in Pahang, amounted to 718,446 acres. As Cant observes, this land was chosen <u>not</u> for potential commercial cultivation, but was poorer quality land suitable for "subsistence and semi-subsistence agriculture" (Cant 1972:115). One effect of this land policy, which continues to curtail development in the Malay sector today, was that Pahang Malay lands were too low in quality, and too small in area, for the rapid expansion of local level indigenous commercial scale agriculture to occur.

While Pesagi informants report that most households began planting rubber just after the 1925 flood, it was many years before rubber sales became a major means of earning cash. Pesagi villagers, attracted by the high rubber price of 1910 (See Appendix G), saw this as a new and easy way to make money. However, laçk of suitable land, scarcity of planting materials, and limited knowledge slowed adoption of the new crop. Moreover, the first harvests in Pesagi occurred at a time when rubber prices continued low for almost 10 years. Informants say that their first sales in 1932 brought only three dollars per *pikul*[3] with total earnings each month at only two or three dollars per household. By 1934 the price paid to Pesagi villages for their rubber crop was five to eight dollars per *pikul*; and the few households with plantings of an acre or more, earned five to ten dollars each month. However, it was not until 1939, when rubber sold at $20 per *pikul*, that prices reached a level high enough for rubber gardening to become a major means of supporting a household.

By the late twenties and thirties Malay villagers in central Pahang had become remarkably skilled practitioners of a highly diversified market and subsistence economy. The major and most dependable crop was rice, though production seldom exceeded local needs. Curiously, the British Administration thoroughly misunderstood and miscalculated the direction that the Pahang Malay economy would take. Believing that rice instead of rubber would become the major cash crop in central Pahang, they responded with the establishment of a government rice mill near Temerloh in 1935 (Corry 1935). The surplus above subsistence needs turned out to be very meager and the government purchased only 120 tons in 1935 (Ibid.), 228 tons in 1938 (Grist 1939), and a record 334 tons in 1939 (Grist 1940). The mill closed shortly thereafter for lack of supply. There may have been a greater surplus than these figures indicate since it is known that much rice was hoarded in fear of floods, and that the Malay were reluctant to sell at $1.80 per *gantang* when the local market price was $2.25 (Corry 1935:529).

During this period, according to informant's reports, Pesagi people increased their involvement with cash crops and marketing to include chickens, buffalo, fruits and vegetables, wild produce, fish, rubber, and coconuts. When rubber prices were low, marketable forest products were gathered, but Pesagi informants say these gradually had to be abandoned due to low prices, and also due to scarcity after most easily reached forest was felled and planted to rubber.[4] Several villagers operated small river boats carrying bananas, coconut, and areca nuts downriver to Pekan, returning with dried fish, salt, sugar, kerosene and palm thatch (*attap*). This river trade attracted Chinese entrepreneurs to operate large commercial boats, transporting goods between the river villages and the main trading stations at Pekan, Lubok Paku, Chenor, and Temerloh.[5]

MALAYSIAN DEVELOPMENT, SINCE 1957

During the past two decades, the Federal Malaysian Government has planted over one million acres of rubber and oil palm in its Pahang development areas at the Jenka Triangle and Pahang Tenggara. In these newly developed areas, called *ranchangan* or "schemes", there are dozens of new communities, each complete with stores, school, mosque, community center, piped water, and homes. Each resettlement family farms about 10 acres of land under supervision of technical advisers, and crops are sold under contract to the developer, the Federal Land Development Authority. Thousands of farmers from every state of Malaysia are moving to these schemes.

About one third of Pesagi's households have moved to *ranchangan*, but all have retained their holdings back in the village. It is not unusual for families to return to the village after having lived several months or years on a scheme, and it is common for scheme families to regularly visit the village and maintain the operation of some of their village economic activities. There are strong social ties between Pesagi and *ranchangan* where Pesagi people have moved, particularly Kampong Sentosa at Bukit Tajau and Jenka Fourteen in the Jenka Triangle. Pesagi villagers frequently visit family and friends in *ranchangan*, and major weddings, funerals, and other social events, bring scheme and village residents together.

Since the government allocates its greatest development support to *ranchangan*, the river villages are not a primary target for government services. Several Pesagi villagers stated that the government in recent years has been ignoring their problems. This allegation was confirmed by interviews with workers in several of the District Offices at Temerloh. According to Department of Agriculture officials, the policy is to carry out an extension service program in every village in the District; but a review of staff visit records shows that Pesagi and other river villages have often been overlooked in recent years. For example, the Medical Office, whose health program requires workers to make regular visits to every village and to implement sanitation projects, claims that all of their time and materials are expended in the more accessible villages located on roads or near town. The Jabatan Haiwan (Government Veterinary Service) said they did not have enough resources to provide full services in remote areas. These service inadequacies, were in every case due in part to an over emphasis given to scheme development and disregard for the special problems of the Pahang Malay.

The presence of development projects has greatly effected the local market system in ways detrimental to the Pahang Malay. This is due largely to the fact that rubber is the major cash crop of most villagers. The village smallholder finds local rubber dealers swamped with large quantities of superior quality development project rubber, and therefore dealers have little incentive to continue buying the small lots of low grade product such as that produced in Pesagi. The result is that the price villagers get for their rubber is depressed 10 to 15 cents per _kati_. At the same time, the rapidly increasing purchasing power of the _ranchangan_ dweller has led to local inflation in the prices of consumer goods. Pesagi villagers are at a severe disadvantage due to prices which are as much as 20 percent higher in towns and markets throughout the central Pahang region.

The Modern Pahang Malay Economy

Today the Pahang Malay, despite problems introduced by government development schemes in their area, maintain a highly flexible economy readily adapted to a wide range of changes in the natural and social environment. Much of what R. G. Cant had to say about the pre-colonial economy is to some degree still characteristic.

> The traditional Malay economy in Pahang represented no ancient and stable adjustment between man and the land. Instead it was perpetually in a state of flux and adaptation; adaptation not to the immediate habitat so much as to the external factors such as war, invasion, piracy, natural calamity, or changing economic opportunities created by contact with foreign traders. Many of these factors had a depressing effect on indigenous agriculture since they resulted in insecurity and discounted the value of long term investment in tree crops or permanent padi fields. They had, however, created a resilience in Malay society which gave it the ability to overcome disaster or adapt more readily to changed conditions (Cant 1972:56).

It is important to note that the use of the term "traditional," which commonly implies a static or unchanging system, in which rules for behavior, beliefs, etc., are passed down without modification from generation to generation, is used incorrectly by Cant. A more appropriate term would be "progressive," or "forward looking," since the indigenous Pahang Malay economy of yesterday and today is continuously changing and adapting.

Pahang Malay farmers have, out of necessity, used diversification as a means of surviving political disruption, crop failure, and an unfavorable market. Application of the term "diversified" to Pahang Malay agriculture is fundamentally different from the notion of "diversified agriculture" described and implemented under current Federal development plans (see Malaysia 1971:13-16). Diversity in Pahang Malay farming occurs at the level of each and every individual household, with individuals participating in the production of many different kinds of crops over a wide range of habitats. Reciprocity of goods and services is frequent between households, with exchange networks extending far beyond local and adjacent villages. Loss of crops, personnel, fluctuations in market prices, and other problems, may greatly stress this widespread system, but failure (ie. famine, or social unrest) is highly unlikely because of reciprocity between discontinuous environmental and social settings. By contrast, diversity in Federally planned projects occurs only on a national scale. Each scheme involves hundreds of contiguous acres, with hundreds of farm families growing exactly the same crop. When failure strikes such a "mono-crop system," the entire region must bear the consequences of having no alternative means of production (cf. Millikan and Hapgood 1967:24,40, 93). Thus, diversity of the kind practiced by the Pahang Malay provides security to the local community, while diversity in the Federal agricultural schemes (while providing some security to the national economy as a whole) brings the risk of production failure to local communities.

Though rubber is the primary cash crop for most Pahang Malay households, other means of survival are maintained because of an expectation, based on experience, that rubber prices are not reliable. As the graph (Figure 5) illustrates, in 1976-77 sudden fluctuations in rubber prices of 5 to 10 cents per *kati* occurred frequently. Informants say that such erratic prices are always the case. Even subsistence crops such as rice are highly unpredictable, as data from the last fifty years illustrates in Figure 6 (from A. O. Report 1927-37; Barnett 1949; Department of Statistics 1975; and Grist 1935, 1939, 1940, 1953).

FIGURE 5 LOCAL MEAN RUBBER PRICES--PESAGI AREA: FEB 1976 - MAR 1977

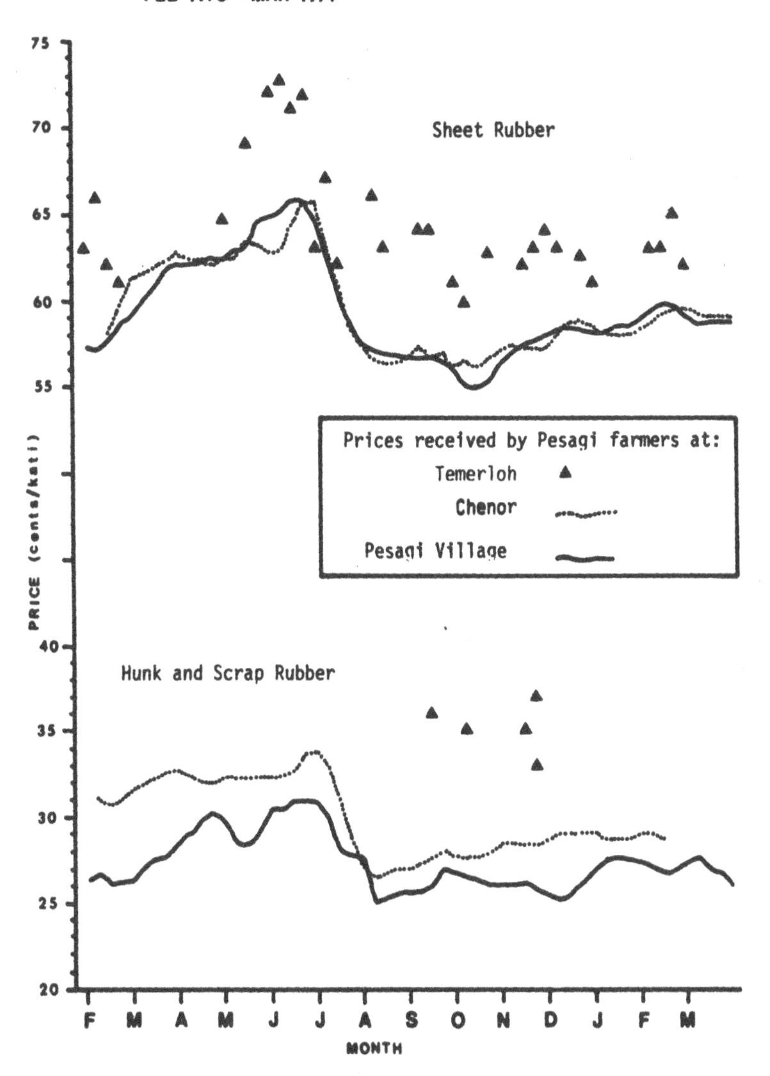

As a precaution, in the event that all main crops fail, each Pahang Malay household has several "fallback" means of survival. Most households maintain a small flock of poultry, a small fruit and vegetable garden, practice a wide range of river and swamp fishing techniques, and a few make extra cash by weaving pandanus mats, selling homemade snackfoods, or taking on occasional wage labor. In case all else fails, every household stores a reserve supply of food in what can be called emergency gardens. The most common plants in these gardens are: manioc (*Manihot esculenta*), yams (*Dioscorea* sp.), sweet potatoes (*Impomea batatas*), and taro (*Colocasia* sp.). These plants are selected because of very important attributes. In each the high energy starchy tuber or corm remains sound and eatable for several months after maturity and can be stored in the unharvested state. Each requires little care after planting except for occasional cutting back of competing weeds. Each grows rapidly and is easily propagated vegetatively, so that a small plot supplies ample planting material for rapid expansion into larger plantings if needed. In each the young leafy shoots can be eaten as green vegetables. When rice is scarce the roots and leaves can also be fed to ducks and chickens. All are non-seasonal perennials capable of producing year-round. While all are occasionally eaten as supplements to the regular rice diet, their stated primary purpose is that of a dependable reserve supply of food.

In summary, the Pahang Malay farm family, due in part to chaotic historical experience over hundreds of years, has evolved a unique and remarkable agricultural system designed to maintain at least a survival level of production under all conditions. Root crops are planted as insurance, in case the rice crop fails. Wild food plants are protected from destruction when clearing and burning garden plots, in case of subsequent cultivated food scarcity. All households maintain the ability to survive for short periods by hunting and gathering. Each individual household is largely a self-sufficient labor force, maintaining a wide range of production skills, so as to insure survival with only minimal reliance on others. These self-reliance measures are augmented by maintaining a wide ranging social network. Moreover, some households are highly transient, moving as necessary between two or more locations, or even dispersing members between several locations, in order to exploit resources in a wide range of locations. These are some of the major features of the indigenous Pahang Malay agricultural system which by inference seem most closely related to their historical past.

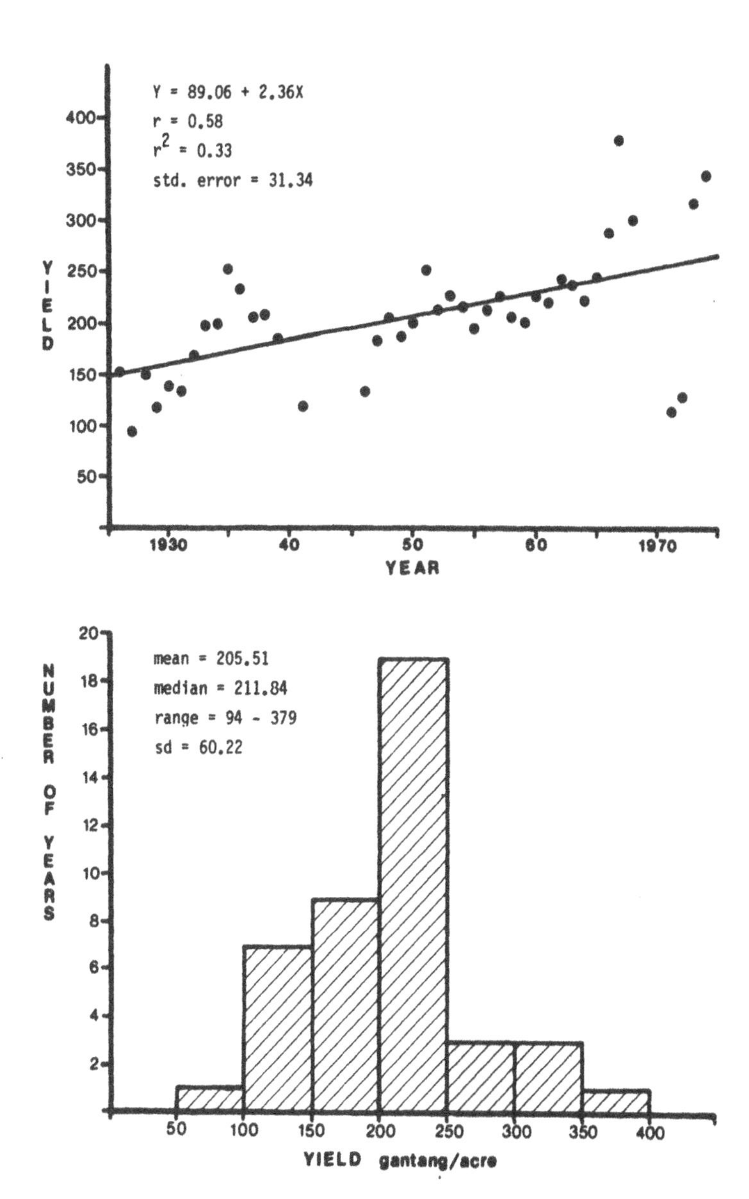

FIGURE 6 PAHANG RICE YIELDS 1925-1975

NOTES

1. The Asli or _Orang Asli_ are those groups of people usually referred to as "Malayan Aborigines." In Pahang they include the Che Wong, Jah Hut, Temok, Semelai, Batek, Temiar, Semai. The most recent general description of these groups is in Carey (1976).

2. This loss of independence for the interior chiefs was a major cause of bitterness against the British, which culminated in an heroic attempt to expel the British and their allies from Pahang in 1891-94 by the Semantan Chief, Dato Bahaman, and the warrior Mat Kilau.

3. A Malay dollar then equaled about 30 cents U.S. currency. For table of measures see Appendix H.

4. Scarcity of some forest products such as rattan, a major material for local construction of houses and fences, caused Pesagi village to request that the District Officer ban collection of rattan for market sale in areas adjacent to their village. Between 1940 and 1950 similar bans were extended to protect many other forest products from sale and to conserve them for local needs. According to villagers the initiative for these bans came from themselves, not from the outside.

5. Boats remembered in Pesagi are the _Ting Wat_, _Haji Jamin_, _Jit Sin_, and _Selamat_.

3
The Natural Setting and Making a Living

THE RIVER

The most prominent feature of the natural environment of Pahang riverine villages is the river. Through central Pahang, the Pahang River's average width is 200 yards, while at Pesagi it spreads another 75 yards to flow around Pesagi Island. Crossing the Mukin of Chenor the Pahang sweeps in a huge bend. Here its main currents cut into the north bank, with the water flowing near Kampong Pesagi on the south bank being sluggish and depositing alluvial materials in constantly changing strands and bars. The average gradient in these central reaches (1.43 feet per mile, Robert Ho 1967:21) is too low to carry heavy sediment in suspension. Hence, the bed is choked with course sand, through which the currents form a shifting set of braided channels.

The landscape and environment are continuously modified by the Pahang River. During high water, sediment is carried over the banks, but instead of being deposited evenly over the inundated areas, the water is too slow once it leaves the river course to carry its load far, and the banks themselves become heightened into natural levees. Thus the leveed riverbanks standing several feet above the swampy flood plain make ideal sites for houses, with fertile alluvium suitable for gardens, good drainage, closeness to water, and positions which catch the cooling river breezes. These levees impede the drainage of side streams into the main river, causing the formation of ponds, which in turn catch sediments washed from surrounding hills, producing swampy areas (paya). On wet lands the natural vegetation is a kind of freshwater swamp forest, which has been cleared by the Pahang Malay in order to cultivate a specialized form of non-irrigated wet rice, called padi paya ('swamp rice'). The surrounding hills were originally covered with lowland tropical rain forest, which in most areas has been removed to establish rubber and other kinds of farms. The accompanying diagrams (Figures 7 and 8) represent Pesagi physiographical features schematically.

The river was busy with boat traffic until World War Two, since it served as the major artery between villages and markets in a trade network extending over 300 miles from coastal ports to the extreme interior regions. At the present time, due to the Pahang road system, the river is little used for transportation and carries only light traffic of villagers going to and from farms, fishing, or to nearby markets.

Pesagi's location, across the river from the main road between Kuantan and Kuala Lumpur, is in many ways a disadvantage. Unlike villages located directly beside roads, Pesagi has no postal service, sick and wounded cannot receive rapid medical care, farmers cannot hire tractors to work fields, the government veterinary service cannot be relied on to treat animals, feed cannot be brought in on a sufficiently regular and dependable basis to maintain large commercial poultry operations, and residents cannot expect the normal range of government development projects to be extended into their village. When buying and selling goods the cost and time lost in high labor intensive river transit has to be added to the high costs of road transport. Sometimes river transit is dangerous or impossible, and Pasagi is isolated from the outside world when dry season water levels are too low for boats, or during the rainy season when the water is swift and debris laden.

Major floods are another problem for riverine villages. The greatest flood in historical times, in 1926, is locally called _Bah Besar_ ('The Great Flood'), and is said by Pesagi villagers to have totally destroyed all houses, fruit trees and cattle. Less destructive major floods have occurred several times during this century, in 1928, 1932, and 1969. Minor floods are beneficial to the riverine villages. These occur more frequently than major floods, sometimes annually or every two or three years. With small floods the silt enriched water improves the fertility of farm land, fills the rice fields, drowns pests and weeds, and strands easily caught fish and wild game in shallow water.

CLIMATE AND ECONOMIC ACTIVITIES

For the most part, western observers have characterized the Malayan climate as _monotonous_ and without major seasonal variation. In B. W. Hodder's book, _Man in Malaya_, it is shown that the monotony of the Malayan

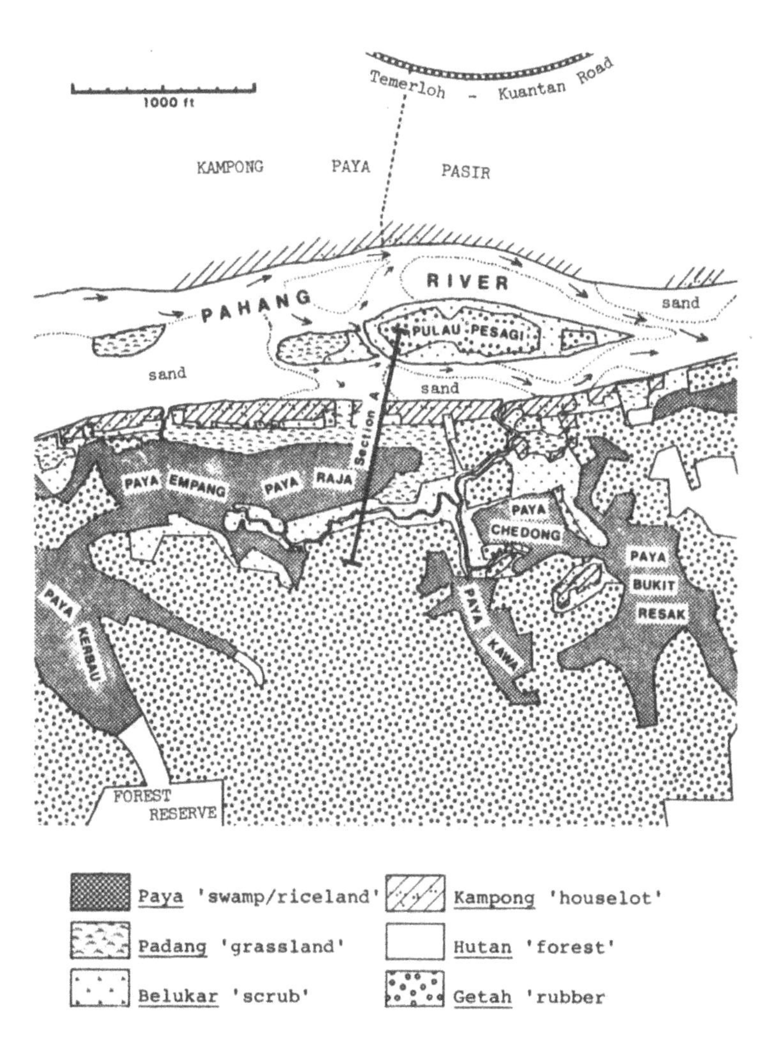

FIGURE 7 PESAGI LAND USE ZONES

climate was a major concern of the colonial government, and westerners feared that the constant high humidity and high temperature was both unhealthy for themselves and lessened the "physical and mental energy" of the local population (Hodder 1959:92). However, as Hodder argues, and my own data supports, what to a westerner is a small or unnoticed diurnal seasonal variation in weather, can be highly significant and easily apparent to a Malay farmer. Social and economic activities are often scheduled and consciously designed to take advantage of a particular kind of rain (a classification is given below). Where the westerner perceives monotony, the Pesagi person perceives great variety. Much of the success of local agriculture depends on economic activities that are well-adapted to small environmental variations to which only the indigenously trained person is aware, or considers significant.

Seasons

Pesagi informants perceive the year as being divided into four periods, each with a different rainfall pattern. In a normal year, beginning in mid-January, there is a sequence of Short Dry, Short Wet, Long Dry, and Long Wet periods. These periods or 'seasons' (musin) are highly irregular and may begin early or late by several weeks, or in some years the seasons may not be distinctly differentiated from each other. The data presented in Figure 9 illustrates this irregularity. For example, 1974 had more rainfall in its dry seasons than in its wet ones. Over the four years, variation between the same season, particularly the Short Wet one, is common.

Informants describe the major economic activities and characteristics of each season as follows:

Short dry. The 'short dry season' (musin panas pendik) lasts about eight weeks, and is expected to begin in mid-January. During this period, rains are few but tend to be heavy, and are interspersed with dry hot periods lasting for many days or several weeks at a time. Villagers see this as a period of rest, following the rice harvest, and turn to more leisurely tasks such as tool repair, houselot tending, and planning for the upcoming rice year. By late March or early April, clearing, burning, and fencing of upland rice nurseries begins. This is the only season when rain is sparse enough that clearing by burning is practical.

Short wet. The 'short wet season' (musin hujan pendik) lasts about eight weeks, and is expected to begin in

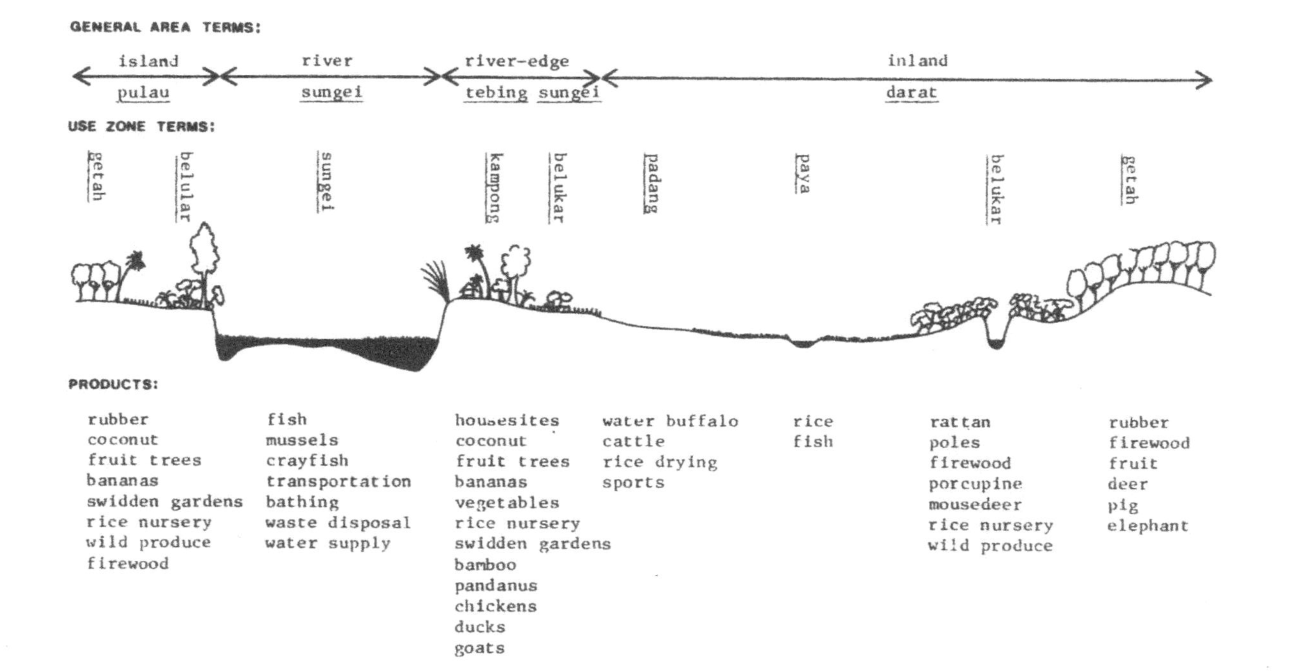

FIG. 8 SCHEMATIC CROSS SECTION (Section A)--LAND ZONES AND MAJOR USES AND PRODUCTS

mid-March. This period has frequent light rains, with occasional heavy downpours which flood the swampy wet lands where rice is planted. During light rains rice seed is planted in the newly burned nurseries. This is the best time for planting vegetable gardens in scrub or on houselot land. The rapid fluctuations of the river makes for good fishing using nets and traps.

Long dry. The 'long dry season' (musin panas panjang) lasts about sixteen weeks, and is expected to begin in mid-May. This period is hot with infrequent heavy rains. Early in the season vegetable crops are plentiful, but become scarce as dryness increases. During long dry spells rubber gardens are weeded by slashing away the competing undergrowth. Firewood is collected, split, and stored beneath houses for fuel during the coming rainy season. Swamp fish are abundant and easily caught because the lowered water level traps them in depressions.

Long wet. The 'long wet season' (musin hujan panjang) lasts about sixteen weeks, and is expected to begin in mid-September. This period has frequent and very heavy rains, which may bring the river to flood stage. Days of continuous rain are spent repairing houses and tools, and making harvest baskets and mats. River fishing is good between rains.

Not specifically mentioned in the above description is rubber tapping, an activity which is relatively continuous year around. Long spells of dryness may significantly decrease the amount of latex which can be tapped, especially during February and March when latex production is already low because of flowering. The trees are more productive during the rainy seasons when latex flow is the greatest, but cannot be tapped daily because wetness causes the latex to coagulate, and spoils the quality of the marketable product. The only time of the year when rubber work is not the primary concern of most households, is the few days (about 37 per year) when rice planting and harvest are in full swing. At other times (about 270) the major work of most households is tapping or marketing their rubber.

Types of Rain

A second feature illustrated by Figure 9 is that the Pesagi labels of 'wet' and 'dry' are not based on the actual quantity of rain which falls in any time period (a datum which none of my informants knew), but rather is based on the incidence of days without rain. Hence, the data shows that the mean number of days

FIGURE 9 RAINFALL AT CHENOR BY SEASON, 1973 TO 1978*

Season:	Year:			
	1973	1974	1975	1976
PANAS PENDIK ('SHORT DRY'): mid-January to mid-March				
Probability of Rain/Day	.167	.317	.267	.067
Mean Inches of Rain/Day	.038	.210	.157	.055
Mean Days Between Rains	8.0	5.9	5.0	11.0
S.D. Days Between Rains	5.8	7.3	3.3	18.5
HUJAN PENDIK ('SHORT RAINY'): mid-March to mid-May				
Probability of Rain/Day	.450	.300	.317	.283
Mean Inches of Rain/Day	.237	.172	.268	.165
Mean Days Between Rains	2.0	2.1	3.9	4.4
S.D. Days Between Rains	1.5	1.7	2.9	7.0
PANAS PANJANG ('LONG DRY'): mid-May to mid-September				
Probability of Rain/Day	.295	.238	.246	.270
Mean Inches of Rain/Day	.273	.151	.140	.268
Mean Days Between Rains	3.6	3.9	3.6	5.2
S.D. Days Between Rains	2.6	3.3	2.6	5.4
HUJAN PANJANG ('LONG RAINY'): mid-September to mid-January				
Probability of Rain/Day	.463	.364	.314	.402
Mean Inches of Rain/Day	.213	.189	.273	.186
Mean Days Between Rains	2.3	2.7	3.6	3.1
S.D. Days Between Rains	1.3	2.3	3.0	3.5
Number of Rains/Year	110	98	96	79
Inches of Rain/Year	79.38	62.59	71.70	66.55

* Data from the Termerloh Irrigation and Drainage Department rain and river returns for Chenor Station.

between rains is greatest for dry seasons, and least for wet seasons. To a Pesagi resident, any three or four days interval without rain is considered as dry and hot. Many plants are adapted to constant high humidity, with roots in the top few inches of soil, and wilt readily if there is no rain for a few days, or they begin dropping leaves if dryness continues for more than a few weeks. Informants frequently pointed out that common plants like _Eupatorium odoratum_, _Clidemia hirta_, and _Melastoma malabathricum_ (see Appendix B and C) show wilting of the leaves after only three or four rainless days, and that rubber trees begin dropping leaves and produce little latex if subjected to about four weeks of dryness. Hence, while heavy rain is very common during 'dry' seasons, the incidence of short dry periods, and the dramatic effect on the environment of even a few dry days, clearly distinguishes for the Pesagi farmer the 'dry' from the 'wet' seasons.

Judging weather correctly is of particular interest to farmers, since the vast bulk of economic activity is dependent upon the weather. This is particularly reflected in the names given to different kinds of rain. For example, one informant described four general types of rain and effects on rubber tapping and other kinds of labor, as follows:

Fine rain. This is called _hujan panas_ (lit. 'hot rain'), _hujan kechik kechik_ ('very light rain'), or _hujan halus_ ('very fine rain').

Description: rain too fine or misty, and too short in duration to wet trees or ground. Falls year around, especially during the hot seasons.

Effect: none on rubber tapping, but hinders work requiring solar radiation such as drying rubber sheets, clothes, fish, or rice.

Short heavy rain. This is usually called _hujan lebat sekejap_ ('short heavy rain').

Description: heavy rain lasting only a few minutes, usually wets only the top canopy of leaves. Falls frequently and year around.

Effect: tapping will pause during rain, and resume if tapping panels are not wetted. Other work stops until rain passes. River fishing with nets (i.e. _jala_) is especially good when these short heavy showers occur at night.

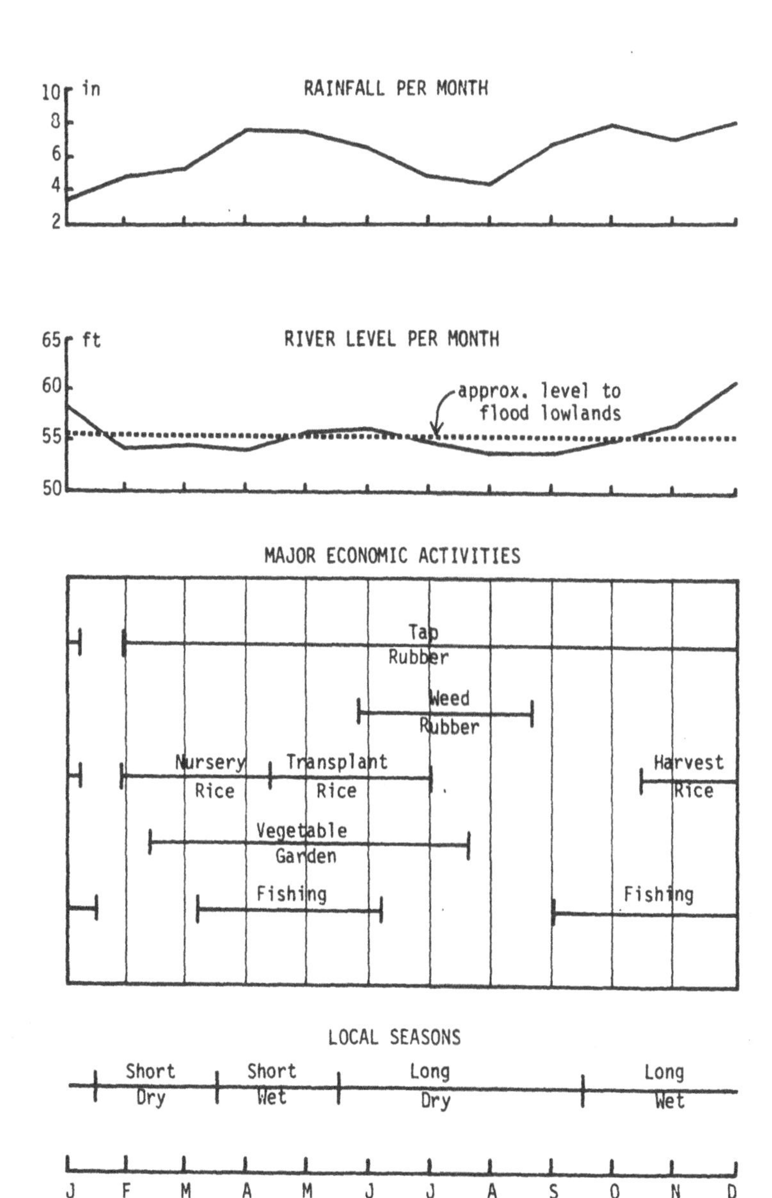

FIGURE 10 RAINFALL, RIVER LEVELS, ECONOMIC ACTIVITES, AND LOCAL SEASONS

Big rain. Names for this type of rain include _hujan lebat_ ('heavy rain'), _hujan besar_ ('big rain'), or _hujan kepal kerayong_ (lit. 'rain of the _kepal kerayong_ fruit') because large quantitites of debris, including this fruit [probably _Parkia javanica_], are washed into waterways and swamps.

Description: very heavy rains dropping 2 to 6 inches of rain in heavy downpours lasting usually less than one hour, bringing local flooding to low ground. Common during short wet season, but may come during any season.

Effect: all kinds of outdoor work impossible. Rubber tapping may stop for 2 or more days until bark dries. Brings one or two days of excellent river fishing with cross nets and traps such as _pengilau_ and _terubing_ (see Appendix D).

Extended big rain. Recognized names are _hujan panjang_ ('long rain'), and _hujan tekujoh_ ('extended rain').

Description: heavy rains lasting several days and dropping several inches per day, usually leading to local or general flooding. Common only during long wet season.

Effect: all outdoor work stops, except for rice harvest which may require use of boats. Villages may be isolated from outside by severe floods.

In order to maintain steady production, each household must shift economic activities as the weather changes. One informant, Mat Nor, reports that sustaining a continuous high level of production requires that a household is always prepared to substitute other productive work when inclement weather interrupts normal work.

> So, what if you plan to tap rubber tomorrow, but at eight or nine o'clock [P.M.] it rains? You can not tap rubber, but you can select some other work. Tomorrow you can go outside [the village], to Temerloh and buy the goods you need...or you can select work...like weeding [rubber or vegetable] gardens,...you can harvest rice or go fishing....If you truly are one who is a successful villager, you cannot waste the day. Whenever you cannot tap rubber there are 10 or 20 other kinds of work to be done.

My own surveys suggest that most households seldom work

to the extent that Mat Nor's statement suggests. Following a good rice harvest or periods of high rubber prices, the average household has an accumulated store of food and cash which enables them to spend several rainy days at more leisurely activities.

THE PESAGI LAND USE AND SUCCESSION PATTERN

Land Classification

When Pesagi residents talk about the natural environment, they commonly use four natural feature categories as classifiers. These are: 'river' (sungai), 'bank' (tebing), 'swamp' (paya), and 'hill' (bukit). In general, any area of Pesagi is named by using one of the four classifiers plus some specific designator, as in:

Tebing Chedong 'the riverbank at Chedong Village'

Paya Raja 'the Rajah's Ricelands'

Bukit Resak 'Resak Hill'(resak, a Shorea sp. tree)

Sungai Pesagi 'The Pesagi River.'

The four specific categories are usually applied as follows:

1. River--natural waterway in which flowing water is usually present.

2. Bank--narrow strip of land bordering a water body or swamp.

3. Swamp--boggy area, or area where shallow water stands. Swamps may include deeper 'holes' (kubang) where water remains even during droughts, or 'channels' (alor) where water flows during heavy rains or floods.

4. Hill--land which generally remains above the level of flood waters. Frequently these areas are called 'high ground' (tanah tinggi).

A second way residents classify areas of the environment is by the type of usage that the area supports, as illustrated in Figure 11.

FIGURE 11
FOLK CLASSIFICATIONS OF LAND AREAS BY USE

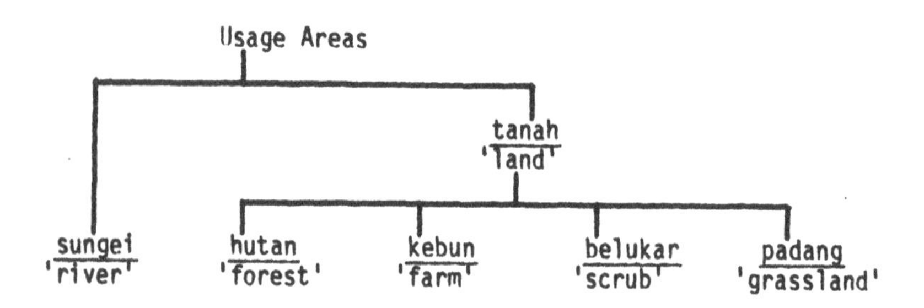

The major uses of each of these areas is as follows:

River. The river is the major source of water for drinking, bathing, and laundry; major disposal area for body wastes, garbage, and offal; area exploited for fish (see Appendix D), prawns, and mussels; transportation facility for moving heavy loads to other parts of village, for traveling to other river villages, and to the riverine market town at Chenor, or to reach the road network on the opposite shore at Paya Pasir.

Forest. The forest is the source of locally used building materials and wild produce; dry lands for conversion to rubber and other gardens; wet lands for conversion to rice fields; area for hunting forest fauna (see Appendix F), and source of commercial timber.

Farm. Farmland is any area from which undergrowth and non-useful trees are cleared as a step in preparation or maintenance of a crop. The farm type of usage occurs on larger river islands, houselot garden space, and productive portions of swamps and hills. In addition to the crops themselves (see Appendix C), fish (see Appendix E) are abundant and easily caught in the alor and kubang of rice fields, firewood is generally available in older rubber gardens, cattle graze on any farm area not protected by fence, and many useful wild plants occur on farmlands (see Appendix C).

Scrub. These are fallow lands recovering from any earlier clearing or farm use. The dominate vegetation includes wild shrubs, herbs, and trees (see Appendix B and C). Where soils are fertile new farms can be established after regrowth of only 5 years, but where soils are poor 10 or more years are required. Healthy scrub forms a rich habitat which may contain fruit trees and other useful plants left from previous farms, attracts game animals, is an abundant source of firewood and some building materials, and also medicinal, food, and other useful wild plants.

Grassland. The grasslands are areas which are primarily open and dominated by grasses forming thick sods, with occasional patches of herbs and shrubs resistant to grazing (see Appendix B and C). Useful mainly as grazing area for buffalo and cattle.

While the plant species common to each of these use areas are somewhat distinctive, there is also a considerable overlap, as demonstrated in Appendix B. Lands which are husbanded in similar ways, or put to similar use, like scrub and rubber farms, share many plants in common. Houselots, which have the greatest diversity of uses, share many plants which are also common to scrub and rubber. The only truly specialized use area, swampland, has the least amount of common plant species.

Land usage areas are ranked from most to least desirable in the order: forest, farm, scrub, and grassland. The ranking reflects productive potential and perceived value. Forest is most desired because conversion to cropland is an inexpensive way for households to increase land holdings. Through such conversion farmers obtain use-rights, and may profit from the harvest of timber. Next in value, is developed farmland planted to long term crops such as fruit trees and rubber, or suitable as rice land. Healthy scrub is only slightly less desired than farmland, as it can be used for a single season vegetable garden and rice nurseries; it can be replanted to long term crops; and it is an area of high species diversity with many useful products. Undesirable is land with grass and weed growth, present where land has been abused by excessive clearing that exposes the soil to the damaging effects of sun and rain, or has been over-grazed by cattle. Grassland is locally considered to be impractical or impossible to convert into cropland, and it provides such meager forage for cattle that it is abandoned and allowed to revert back to public or government property.

The most intensively used areas before cash crops were introduced in the early twentieth century were the swamps and their immediate borders. The well managed *paya* provided fish, and a wide range of niches for crops such as rice and taro. The bordering drier land was used for swidden gardens of yams, sweet potatoes, bananas, tapioca, and millet. Today, only rice remains as a *paya* crop, since the drier border areas have been over-exploited as garden sites and as collection areas for building materials and firewood, which together have caused an invasion of relatively useless grass.

Over time, natural forces may drastically change the area of land available for different types of usage. Large floods bring permanent changes to land use patterns, as low-lying areas are covered by several inches or feet of silt and sand. Many acres of Pesagi which were once swampy are now dry, and in one case some 45 acres has been covered by deep water due to silt blocking drainage channels. The evidence, in the form of field boundaries and informant's testimony, is that about 426 acres have at some time been used to grow rice[1], but today only 96 acres with suitable elevation and contours remain (see Figure 12). This is a reduction to only 22 per cent of the area which at some time in the past was suitable for rice. About 278 acres of this wetland is presently covered by grass. Similar reductions in rice areas have been recorded in other parts of Pahang, with silt damage after floods of 1926 and 1932 destroying more than a third of a 1,000 acre rice area at Pulau Tawar in the Lipis District (A.O. 1937:49; Birkinshaw 1941:29).

Succession Patterns

The introduction of cash crops, and the transformation of the economy to a cash orientation, have brought about radical changes to Pesagi land use patterns. Sporadic high rubber prices encourage the planting of rubber on any available land, including forest, scrub, garden plots, and houselots. Sporadic low rubber prices force families to fall back on subsistence food production. As the hill forest and other lands are converted into rubber farms the availability of land for food and other cash crops decreases. Grass becomes an increasing problem as land is used more intensively. Over the last 50 years, the environment has been changed to the point that the land closest to villages is in various stages of scrub regrowth, and increasingly supports nothing but grass and weeds. This reduces the area available for vegetable and fruit gardens close to housesites. Land is not available further away because of rubber gardening. Forest land, where it is not already used up, is too far from present village sites, and is under government restrictions on use by villagers.

Observation suggests that the indigenous Pahang agriculture system brings about a dynamic succession of land conditions (cf. Bartlett 1956; Conklin 1967) through which much of the land repeatedly cycles.

FIGURE 12 PESAGI LAND TYPES AND AREA IN ACRES

GENERAL LAND TYPES		CROP LANDS	GRAZING LANDS	LIGHT USE	TOTAL AREA
RESIDENCE AREA	(# of Households)				
KAMPONG PESAGI	(56)				
Houselots		22	7	15	44
KAMPONG CHEDONG	(20)				
Houselots		11	7	9	27
KAMPONG BUKIT RESAK	(14)				
Houselots		17	0	15	32
KAMPONG LOMPAT	(19)				
Houselots		20	15	30	65
K. TANJONG BERANGAN	(31)				
Houselots		39	5	13	57
TOTAL RESIDENCE	(140)	109	34	82	225
FARM LAND					
PAYA AND RICE					
Greater Pesagi Paya		40	246	4	290
Paya Chedong		18	3	1	22
Paya Londang		4	4	0	8
Paya Kawa		14	1	1	16
Paya Bukit Resak		10	15	45	70
Paya Lompat		10	9	1	20
TOTAL PAYA AND RICE		96	278	52	426
RUBBER					
Granted		440	20	640	1100
Non-granted		200	0	40	240
TOTAL RUBBER		640	20	680	1340
OTHER FARM					
Rice Nursery Reserve		0	3	26	29
Pesagi Island		25	5	14	44
TOTAL OTHER FARM		25	8	40	73
TOTAL FARM LAND		761	306	772	1839
GOVERNMENT LAND					
Forest Reserve		0	0	240	240
Grasslands		0	110	0	110
Graveyard and Mosque		2	10	3	15
School Lands		0	9	2	11
TOTAL GOVERNMENT LAND		2	129	245	376
TOTAL PESAGI ACREAGE		872	469	1099	2440

FIGURE 13
LAND USE SUCCESSION PATTERN

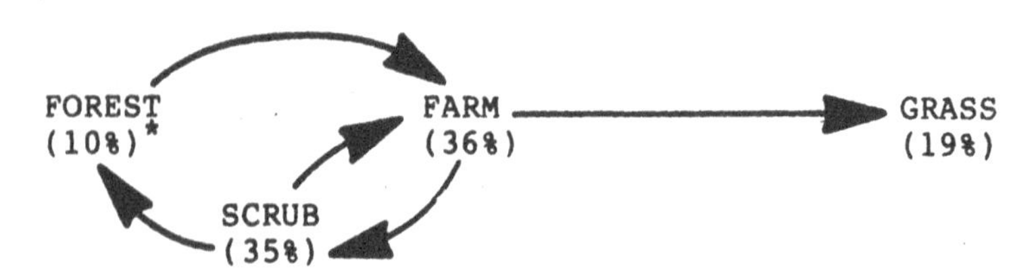

*percentage of Pesagi land in each condition in 1976.

These cycles can best be expressed in diagrammatic form as shown in Figure 13. The arrows in the figure refer to land condition changes through time.

Grassland, the least productive land condition, is relatively terminal or permanent. Two ideal succession cycles continue to return land for farm use without losses to grass. One can be called the "long fallow cycle," and involves a continuous cycle of FOREST to FARM to SCRUB to FOREST. This long cycle requires an undisturbed fallow period of up to 60 years in central Pahang (Ho 1967:29), but can no longer be practiced due to land shortage. The other way in which farmland is continuously brought into production, is the use of the "short fallow cycle," which re-uses scrubland before it returns to forest with a continuous cycle of FARM to SCRUB to FARM. The short cycle requires a fallow of only 5 to 10 years. However, if the crop planted is a tree crop, like rubber or fruit, the crop occupies the site for many decades, perhaps 30 to 50 years, and if long-term crops are planted on a major portion of the available land, then land available for short-term uses, particularly for locally consumed food crops becomes scarce.

It is illustrative to compare the land use patterns of Pesagi with two other similar Pahang Malay areas. Data is available from a study by Robert Ho (1967) of an area near Temerloh comprised of parts of Mukim Kerdau, Sanggang, Songsang, and Bangau, and from the study by Rudolph Wikkramatileke (1958) in the Pekan District of the mukim of Pulau Rusa. Both of these areas, like Pesagi, have a similar history of subsistence rice production, and began planting rubber, coconut, and other cash crops in the first few decades of this century. The cultural descriptions by the authors show that all three areas share the same agricultural technology, and have similar values regarding the

FIGURE 14
ACRES PER PERSON BY SUCCESSION CONDITION

	FOREST a/p* %	FARM a/p %	SCRUB a/p %	GRASS a/p %	TOTAL a/p %
P. Rusah	5.9 (73)	1.7 (20)	.2 (2)	.4 (4)	8.1 (100)
Temerloh	1.1 (24)	3.1 (66)	.3 (6)	.2 (4)	4.7 (100)
Pesagi	.4 (10)	1.5 (36)	1.5 (35)	.8 (19)	4.2 (100)

*acres per person

desirable "target conditions" for the lands they use Because of the similarity of the three areas, and because each represents a different time period, it is assumed that with some caution the data is representative of the development sequence for the entire central Pahang area over the last 25 years. Harold C. Conklin (1967:110) maintains that an easy way "to assess the relative economic standing of an Ifugao agricultural district is to note the degree to which ...[a] target condition [of desired land types] has been approached." Simply stated, the target condition for the Pahang Malay is to have some forest area available for hunting and gathering, to have an adequate area of productive farmland, and to avoid the creation of relatively useless grassland.

The comparison of these three areas illustrates not only the land succession pattern, but also the integration of cash cropping into indigenous agriculture. All three areas have rapidly, since the 1926 flood, converted forest land into cash crop farms, mostly rubber. This is shown in the first two columns of Figure 14. Temerloh appears to have reached a very enviable situation in the mid-1960s, whereby the target condition was met to a much greater degree than the other two areas. However, my own inspection of these Temerloh area villages in 1975-76 shows that the 1960s condition was transitional, and that the high proportion of farmland (i.e. 66%) could not be maintained because it was largely comprised of old and increasingly non-productive rubber trees. Pesagi had a similar high peak of farm land some ten to fifteen years ago, as probably did many other Pahang Malay villages. Fortunately for these villages, this period of great farm area coincided with a period of high rubber prices (see Appendix G; (cf. Courtenay 1956: 106-112; McHale 1967), particularly during the Korean

War, which brought about ten years of local well-being.[2] Today the situation elsewhere is more like that of Pesagi, where shrinkage of farmland area has come about as resources recycle and balance is restored between fallow scrub and productive farmland.

Many Pahang Malay farmers have adopted methods of farming, shown by generations of indigenous experience to lead to recycling of resources, in order to insure that their farms become healthy replantable fallow scrub rather than next-to-useless grassland. The basic principle of the indigenous method is to create highly diversified mixed gardens, either by integrating gardens into existing scrub, or by encouraging scrub to grow between fruit and rubber trees (see Figure 15). This contrasts with the "modern" method, in which only a single crop is permitted to flourish and other plants are deliberately eradicated. The indigenous method ensures a moderate though sustained level of yield, and a continuous area of healthy productive land. "Modern" methods, in contrast, involve a risk that after initial peak yields are reached, productivity will decrease indefinately or irrecoverably. As Figure 15 illustrates, an indigenous garden integrates rubber, fruit trees, vegetable plots, foraging area for poultry and cattle, and rice nurseries, all within very small patches in an area where fallow regrowth is the predominant vegetation. Hence, as intensive use of each garden plot drops off after one or two plantings, natural succession rapidly heals the spot with regenerative growth.

The Pesagi farmer, when establishing a new garden or farm on forest or scrub, utilizes only partial and selective clearing, leaving useful species undisturbed, with the purpose of diminishing the intensity of natural growth rather than removing it.[3] After the crop is planted, the undergrowth is pruned back selectively to give the most useful wild plants, and the planted crop, an advantage over plants of lesser utility. Following several years of this kind of cultivation the benefits of the indigenous growing method are readily apparent. First, rubber (or other) gardens become areas rich in useful resources, among which are many well-adapted naturally occurring species. Second, the value of the resulting enriched mixed-resource garden may exceed that of mono-cropping the same area.[4] Third, famine becomes highly unlikely even when rubber prices are very low, because of the continuous availability of a great range of useful products. And fourth, environmental damage is avoided.

A major shortcoming in Pahang Malay villages is that not all farmland is recycled for future farm use. Some farmers, influenced by "modern" methods, clear all undergrowth from gardens to increase yields. Because

FIGURE 15 MIXED GARDEN: HH5 JUNE 1976

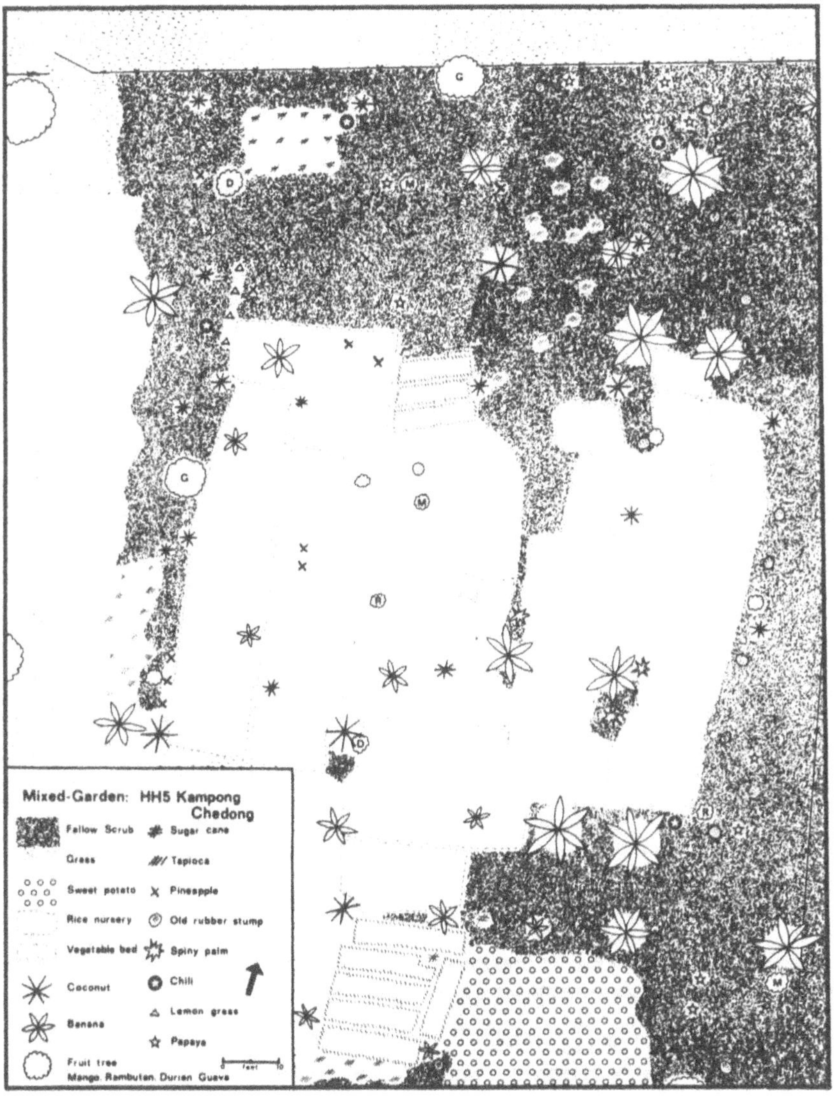

of poverty, many have been unable to afford fencing to keep cattle out of their gardens. In general, the cattle population is becoming critically large; and cattle have in fact become the major pests attacking crops and hindering fallow growth. Government programs to distribute cattle to villagers help large landowners but contribute to the poverty of poorer farmers.

LAND USE AND OWNERSHIP

It is customary to acquire "use-rights" to property by either pioneering the development of forest for farm use, which establishes "founder's rights," or by inheriting a "use-right" from some previous "founder." Any individual simply retains use-rights to a specific site as long as he continues to use it. The length of usage depends on the fertility of the soil and the nature of the crop. A rice nursery, for example, is used for one season only, since the rice itself is transplanted within two months, and other crops planted at nursery sites are of short-term duration such as corn, beans, and pumpkins. A clearing to plant medium-term crops like bananas, tapioca, and papaya may have a "user" for 5 to 7 years; while a clearing for long-term crops like coconut, durian, or rubber, is used for so long, at least 40 years, that it is seen as having a relatively permanent "user." In the past, when forest was abundantly available, cash crops were unknown, and the customary land use system was very closely related to production, since land had no social or economic value other than that conferred by the products it produced. No one could "own" more land than they could actually place in production, and the land resources of an area were rather evenly distributed and recycled among users over time.

Today, the customary land use system still pertains, but it is being replaced rapidly by an introduced system of "legal ownership." The "legal ownership" of real estate began with the British administration,as a means to regulate land use and generate taxes. From the government's perspective, then and now, only long-term usage lands can be alienated. Today only part of village lands have legal "owners," and therefore, fall under the jurisdiction of the Lands and Survey Department, which conducts surveys, issues deeds approving usage, and assesses annual taxes. These "owned" lands (tanah geran, lit. 'land with grant') include most riceland, houselots, rubber, and fruit orchards.

Land under short-term or medium-term crops has no legal owner. While customary use-rights are recognized locally, and informants say that specific sites belong

FIGURE 16
MEAN DISTRIBUTION OF LAND PER HOUSEHOLD

Usage Area	Total Acres	Mean Acres per HH	Income Range in Percent
1. Rubber gardens	640	4.57	40 - 90
2. Rice fields	96	.69	15 - 50
3. Other crops[a]	136	.97	10 - 30
4. Regular light use[b]	859	6.14	15 - 25
5. Occasional use[c]	240	1.71	0 - 10
6. Grazing land	469	3.35	0 - 5
	2440	17.42	

[a]These other crops include coconuts, vegetables, fruit, and poultry, which utilizes houselots, and Pesagi Island.
[b]Light use activities include collection of firewood, building materials, and wild produce from swidden and old rubber gardens, and from the borders of fields and rivers.
[c]Occasional activities include hunting, fishing, and collecting wild products on forest lands.

to specific users, the same locations are also designated as 'government land' (<u>tanah kerajaan</u>), and sometimes as 'forbidden land' (<u>tanah haram</u>). This land in general includes all forest, and any illegally used lands on which people have built houses or established farms. There is also a small area of 'government land' which is for public use such as cemeteries, drainage ditches and dams, river margins, some rice nursery reserves, most grasslands, school and mosque property, reserved road right-of-ways, and lots with Government buildings like the Birth Clinic and Community Hall.

To a large extent the greatest measure of social and economic success for the agricultural based Pahang Malay is ownership or access to enough land to meet production needs. The members of a successful household, as population continually expands, experience increased difficulty in obtaining adequate rice, fruit orchard, swidden garden, house, forage, and other land to meet expanding needs. Figure 16 divides land into six resource areas which are required by most households, together with the mean acreage of each usage area per household. The percentage of income range indicates the relative importance of each type of usage area, and also that a household must have access to a range of usage areas in order to maintain an adequate income. No household can service without either rubber or rice, and the strongest position is a combination of both. A household with only the mean acreage of all six usage areas, could be expected to range from slightly below its income requirements to a considerable excess above.

Of the almost one hundred households surveyed, the disproportionate distribution of land between households is readily apparent. For lands producing cash crops, such as fruit and rubber, there are three households which have 10 times, 10 which have five times, and about 20 households which have two times the mean acreage.

Parcels of land are constantly being divided, as junior household members marry and set up independent households. Traditionally, division is supposed to apportion the property evenly among one's children. However, since an estate is parceled out at different times, and since the need of the recipient is often taken into consideration, some reciprients get more than others. As parents age and require less land to meet their own needs, they distribute it to their offspring. Most land is already in use by the younger generation long before old age causes the elder couple to retire from working. This method of land distribution is equitable only when land is available in surplus. Otherwise, older siblings who need land to support families exhaust the supply long before their younger brothers and sisters reach marriageable age. This results in a pattern in which the eldest members of each generation marry young, establish large families, and stay in Pesagi as farmers, while the younger members are more likely to continue in school, partly supported by elder siblings, and to eventually find employment elswhere.

Ideally, a household's land holdings are large enough that when divided among succeeding children, the shares equal at least one half the area required by a new household (or a portion large enough to support a small family when worked together with a spouse's share). In order to perpetuate itself from one generation to the next, every household must expand its holdings by a a factor of one half its total number of surviving children. Such expansion has not been possible in recent decades, and the result has been that only one or two children from each household can marry and settle in Pesagi.

The land shortage problem is alleviated in part by share-cropping, renting, and charity; but these solutions introduce their own problems. At least 20 Pesagi households regularly share-tap rubber. More than half the households planting rice in 1976 rented land from external landlords (primarily the Chenor Royalty). Many households allow their poorer friends and relatives to farm excess land. If these land sharing mechanisms work perfectly, most households in Pesagi could live comfortably, but such is not the case. Many acres of good Pesagi farmland are unproductive for lack of users. The reasons are primarily social. Charity is not given widely, but only to a select few. Share-cropping is so socially demeaning that many choose to remain poor. Persons working for shares are usually landless and occupy the lowest social rank in society. Hoarding property, moreover, is one marker of high social standing. As a result, the few wealthy people tend to acquire more land than they need or can use, and the poor tend to farm meager holdings at ever increasing intensities. The disproportionate allocation of land drives the land use succession pattern towards decreased production (i.e. poor scrub grass) in a disproportionate way, that is, poor villager's land is ruined at a faster rate than the land of rich villagers. This benefits the rich in two ways. First, the poor become destitute and are increasingly forced into share-cropping; or they are forced to sell their landholdings to more wealthy villagers. Secondly, each farm lost to grass provides increased grazing area for cattle, which are primarily raised only by the village elite. These problems are all related to population density and shortages of land, and they suggest that the village way of life, and indigenous agriculture, is in serious danger of extinction from land shortage. This is made all the more serious, by the fact that plantation and government-operated large-scale agricultural development programs are in trouble as well.

NOTES

1. It is very unlikely that such a large number of acres were ever planted at the same time. The early form of wet rice agriculture was swampland swidden, requiring large areas of fallow. Also, the variation in water level from year to year affects both the location and area of swampland suitable for planting at any one time.

2. Robert Ho's (1967) work shows the income per household per month 10 years ago to have been $154 in villages near Temerloh. The average Pesagi household today earns about $95. Indicators of an earlier period of wealth remain today in the form of grand houses, diesel powered river boats, rice mills, water pumps, and electric generators. Since the late 1960's such

expensive items have been beyond the means of most households.

3. In the use of *paya* for growing rice by indigenous methods of rice farming the natural species preserved are a diverse range of aquatic plants and animals. Fish are a major naturally occurring resource and a major local food resource, but generally cannot adapt to modern technological inputs such as water control, fertilizers, pesticides, or other agricultural chemicals.

4. This argument, though not objectively substantiated with the actual measured value of products, is so frequently made by informants that it is obvious that it is widely believed and possibly true. Families who have moved to schemes claim that cash earnings during periods of normal prices and yields, because of inflated prices of consumer goods available to scheme families, make them actually less well off than if they had stayed in the village. This is because village agriculture supplies free or very inexpensive building materials, fish and game, firewood, and a wide variety of foods and other necessities.

4 Rice Varieties

In Pesagi over 40 rice cultivars are recognized and planted. Visits to other villages and encounters with Malay from other parts of central Pahang, suggests that there are at least one hundred locally recognized rice cultivars. As will be demonstrated below, a major characteristic of these cultivars, or at least those recognized in Pesagi, is that each is quite distinct from the others in terms of readily recognizable morphological characters and habitat requirements. Rice interbreeds readily, so that rice farmers must utilize highly effective skills of seed selection in order to maintain the characters of each distinct cultivar over time. Some writers, such as Grist (cited below), have argued that a major reason rice farmers have developed highly elaborate seed selection skills is for aesthetic reasons. As far as Pesagi is concerned, my own data suggests, quite the contrary, that the large number of rice and some other cultivars is related to the great range in small micro-habitats, to the irregularity of weather, to the inability to control water supplies, and to other vagaries in growing conditions. Were it not for a wide range of specialized cultivars, the area over which any crop could be grown would not be great, and agriculture would be an unreliable means of earning a living much of the time. For the purpose of discussing the importance of these many cultivars in Pahang agriculture, I have divided this section into three topics: classification, environmental considerations, and adaptation processes.

RICE CULTIVAR CLASSIFICATION

Pesagi farmers use a number of easily recognizable characters to distinguish between cultivars, with those which are most commonly encountered being presented in

FIGURE 17 LOCAL RICE CULTIVARS

GANGSA (vitreous):
1. *gangsa puteh [8]
2. gangsa puteh alus
3. gangsa puteh kasar
4. *gangsa puteh ekor musang [21]
5. gangsa anak musang
6. *gangsa darah belut [23]
7. gangsa merah
8. gangsa tembeling
9. *seri bomi [7]
10. gading kedah
11. *jambai [20]
12. *ekar kerbau [11]
13. *serendah
14. *serendah tinggi
15. serendah rendah
16. *serendah choring [9]
17. *serendah kuning [10]
18. serendah merah
19. *serendah puteh [17]
20. *serendah borek
21. *mele
22. *mele kuning [18]
23. *mele puteh
24. *mele merah [1]
25. *mele choring

LEMBUT (viscous):
1. *lembut nunget
2. *lembut langit
3. lembut mayang
4. lembut jamban
5. lembut nawit
6. lembut kerbau
7. lembut kumpai
8. *terong lembut [3]

PULUT (glutinous):
1. *pulut hitam [4]
2. *pulut merah
3. *pulut puteh
4. *pulut dukong [6]
5. *pulut kembang merdu [2]
6. pulut kembang merdu puteh
7. pulut kembang merdu merah
8. *pulut bunga machang [5]
9. pulut janda kaya
10. *pulut julai [15]
11. *pulut naga nulu [19]
12. *pulut ayun
13. *pulut gemuk [22]
14. pulut godok
15. pulut genik
16. *pulut dadeh
17. pulut gading
18. *pulut siam

---

* Varieties planted in 1976-77.

[#] Specimen numbers for collected samples.

Figure 18. Special importance is given to the hardness or texture of the grain, since this determines its use. There are 23 cultivars in the general category of *gangsa* or *padi biasa* ('common rice'), which have hard starchy grains that are dry and "fluffy" when cooked (see Figure 17). These are frequently referred to as "vitreous" in English because the uncooked grain is translucent; or as "common" because these are the most widely cultivated on a worldwide basis. Rice varieties of the vitreous type are the major food grown, eaten and, when necessary, purchased by Pesagi households (the per capita consumption for adults being about 278 pounds per year).

There are 18 cultivars in the category called *pulut* ('sticky'), having soft grains that become sticky and clump together when cooked. In the past, these were only required in small quantities for consumption during special ceremonies, for making rice-starch cosmetic and medicinal preparations, and for making traditional snack foods and cakes. In recent years *pulut* rices have commanded a premium price in local market towns, since they symbolize high status and are desired by urbanized Malaysians. While Pesagi growers formerly considered *pulut* only a minor, locally consumed crop, they are increasingly aware of the cash crop possibilities of these varieties.

The final eight cultivars are called *lembut* ('soft'), described by informants as so soft and almost gelatinous when cooked that the label "viscous" is perhaps appropriate. *Pulut* and *lembut* have not been distinguished as separate categories by western rice specialists. The *lembut* cultivars are pounded into flour and used for making a special class of cakes and pastries,[1] though unlike the cakes that require *pulut* rice flour, informants say that in most cases imported wheat flour is an acceptable and cheap substitute for *lembut* flour. In 1976 there was little local interest in the *lembut* group, as evidenced by the fact that only a couple of households planted small plots to three types.

Color is another important character, and while almost any part of the rice plant can be pigmented, it is specifically the color of the husk and the pericarp or outer layers of the kernel that merit significant cultural attention. Colors noted are *puteh* ('white'), *kuning* ('yellor to light brown'), *merah* ('dark brown to red'), and *hitam* ('black or any dark color'). A color term is often part of the cultivar name, and it serves to distinguish closely related cultivars, such as *mele kuning* and *mele merah*. Some have fanciful names which are indicative of color, such as the *ganga* with red-brown husks called *darah belut* ('swamp-eel blood') and the *gangsa* with black husks called *ekar musang* ('civit-cat tail').[2]

Other terms are descriptive of husk color patterns, such as _choring_ which describes splotches and stripes of red and white on a _serendah_, and _bunga machang_ ('machang flower') glutinous rice which to the local eye resembles the striking red and white flower of the _machang_ variety of mango.

While vitreous rice may have husks of any color, the pericarps are always white. This is probably a local cultural preference, since colored vitreous rices are common elsewhere (e.g., in Sarawak and the Philippines from personal observation). When buying "food" rice, Pahang people demand white polished types, and they boast about the whiteness of the rices they grow themselves.[3]

The attitude towards color in glutinous rices is quite different, with the full range of colors being in demand. Each type and color of glutinous rice has specific ritual uses, or is required for making special confections or cakes.

Growers frequently distinguish cultivars according to the size and shape of the grain. Terms describing size and shape include: _panjang_ ('long'), _pendik_ ('short'), _gemuk_ ('fat') and _bulat_ ('round'). Informant's shape labels for spikelets (i.e. grain in husk) are given in Figure 18 together with actual measurements in millimeters. It is apparent that any rice over 8.5 millimeters in length is called 'long,' while all others are 'short.' A further division can be made according to the ratio of width to length. Grains that are one third as wide as they are long are called 'fat,' and those where width approaches half the length are called 'round.'

The characters just described above -- that is color, shape, and texture -- are those most frequently used in Pesagi to identify a specific cultivar. In a series of identification tasks in which informants were asked to identify a set of 18 cultivar samples, it was discovered that the process utilized consists of a logically ordered set of procedures based on these three characters. An informant, who attempted to teach me how to identify rice, emphasized the following steps:

Step 1: Determine which rice heads are short-grained and which are long-grained.

Method 1: Observe and compare spikelet lengths.

Method 2: As an alternative, or test of divisions made by method one, ask about water depth (it is common practice

FIGURE 18 CHARACTERS OF PESAGI RICE CULTIVARS

SYMBOL	LOCAL NAME	WATER DEPTH	TEXTURE	SHAPE LENGTH	SHAPE L/W	SHAPE NAME	COLOR HUSK	COLOR PERICARP
JA	Jambai	deep	vitreous	9.0	3.6	long	white	white
SB	Seri Bomi	"	"	9.0	3.1	"	yellow	white
EK	Ekar Kerbau	"	"	9.0	3.1	"	yellow	white
GP	Gangsa Puteh	"	"	8.8	3.0	"	white	white
GP/DB	Gangsa P. Darah Belut	"	"	10.0	3.1	"	red-brown	white
GP/EM	Gangsa P. Ekar Musang	"	"	9.0	3.0	"	black	white
SE/P	Serendah Puteh	shallow	"	8.2	2.7	short	white	white
SE/K	Serendah Kuning	"	"	8.1	2.4	"	yellow	white
SE/C	Serendah Choring	"	"	8.0	2.3	"	wt/rd(1)	white
ME/K	Mele Kuning	"	"	8.0	2.4	"	yel-brown	white
ME/M	Mele Merah	"	"	7.9	2.3	"	red-browm	white
PH	Pulut Hitam	semi-dry	glutinous	10.3	4.3	long	black	black
PK	Pulut Kembang Merdu	"	"	9.9	3.8	"	grey	red
PN	Pulut Naga Nulu	"	"	10.9	3.6	"	lt. brown	white
PB	Pulut Bunga Machang	"	"	9.0	3.6	"	wt/rd(2)	white
PJ	Pulut Julai	"	"	9.0	3.0	fat	yellow	white
PD	Pulut Dukong	"	"	9.0	2.5	"	white	white
PG	Pulut Gemuk	"	"	9.1	2.4	"	white	white
TL	Terong Lembut	"	viscous	6.2	2.0	round	white	white

SYMBOL: Used on Map 5.1 to show planting pattern.
WATER DEPTH: deep = greater than 1 foot; shallow = less than 1 foot;
semi-dry = occasionally flooded.
SHAPE: measurement of spikelets (i.e. grain in hush) in millimeters, with name given by informants.

1. Piebald effect with white and red splotches.
2. The third of the husk near the pedicel is white, with remainder being red.

to grow short-grained cultivars in shallow water).

<u>Step 2</u>: Determine which long grained heads are vitreous and which are glutinous.

<u>Method 1</u>: Remove husk and check for chalky-white opaque (glutinous) or clear (vitreous) kernels.

<u>Method 2</u>: As an alternative, or test of divisions made by method one, ask about water depth (usually, but not always, glutinous cultivars are grown under semi-dry conditions; vitreous are always grown in deep water).

At this point the samples are in four groups, those which are:

(a) Long-grained and vitreous (deep-water);
(b) Long-grained and glutinous (semi-dry);
(c) Fat-grained and glutinous (semi-dry);
(d) Short-grained and vitreous (shallow-water).

<u>Step 3</u>: Determine color of husk and husk color pattern if present. Apply cultivar names to each sample which can be clearly discriminated.

At this point all informants had clearly isolated and named most of the cultivars. However, there are a few pairs, for example <u>ekar kerbau</u> and <u>gangsa puteh</u>, or <u>pulut dukong</u> and <u>pulut gemuk</u>, which have such similar characters, that an additional piece of information is required.

<u>Step 4</u>: Ask who grew it?

More competent informants were usually able to positively identify any cultivar by knowing the source household, particularly if this household was that of an intimate, relative, or near neighbor. Apparently, this is due to personal familiarity with the cultivars grown by members of their own close personal network, and of other individuals living nearby. Additionally, knowing the grower often provided extra clues about the habitat.

Consulting the literature on rice identification shows that Pahang's rice cultivars are so diverse as to include many of the characters known for rices spread over a wide geographical range. However, there are serious problems with trying to relate scientific rice classification systems with the system employed at Pesagi, because one uses gross morphology while the other is increasingly using characters requiring microscopic or chemical analysis, and because there is no universally accepted means of classifying rice. Some experts caution that many variables are accounted for by environmental differences, hence classification schemes are inaccurate (Purseglove 1972:166). In Figure 19 the major classification schemes, based on agricultural, and gross morphological characters, as summarized by Purseglove (Ibid.), are presented together with a table of Pesagi rices. This demonstrates that Pesagi rices are extremely diverse in terms of exhibiting almost the full range of characters used by all classification schemes. A reasonable conclusion is, that comparatively speaking, all Pesagi cultivars are late-maturing varieties, but as a group are otherwise diversified to the greatest possible extent.

ENVIRONMENTAL CONCERNS

The paya ('swamp') soil surface is bowl-shaped when viewed in cross-section (Figure 8), with the water shallow near the edges and deep toward the center. While the Pesagi rice grower recognizes numerous differences in soil, pest problems, and so on, at the most general level the swampfield is described as having four distinct habitats, indicated by the mean water depth during the planting and growing season. The first, the outermost contour at the swamp edge, varies from completely dry during the occasional periods when rain does not fall for several weeks, to flooding up to six inches when it does rain. This "semi-dry" habitat informants call paya kering ('dry swamp'). The second contour begins at the level where water is almost always present, extends to a water depth of about 12 inches, and is called paya tohor ('shallow swamp'). The third contour extends from the shallow area to the maximum depth possible to plant (i.e. an arm's length) and is called paya dalam ('deep swamp). The fourth contour is deep water, if any, at the center of the paya, and includes kubang ('holes') and alor ('channels').

Each individual household's rice farm cuts across several contours and other recognized natural habitats, which form the basis for dividing the planting area into plots suitable to the requirements of different cultivars. Due to habitat diversity, the mean number of cultivars planted per household is three or four, and five or six is not uncommon. Even though two households may have similar or adjacent rice farms suited to an identical set of cultivars, the abundance of cultivars and personal preferences usually cause each to choose differently. The map of Paya Pesagi (Figure 20) illustrates the general location of these natural contours, the boundaries of individual farms, and the general location to which each cultivar or variety was planted in 1976.

Given that Pesagi farmers have almost no control over water, the cultivars utilized must be tolerant to the range of water levels normal to the contour where they are planted. This range can be estimated to vary from the 'dry swamp' contour level by as much as plus or minus three feet, given normal patterns of rainfall.

Water depth tolerance is much greater for long-grained than for the short-grained cultivars. For example, cultivars in the long-grained glutinous group, which are planted to the semi-dry outer contour, are claimed by older informants to have been grown in earlier times on upland swiddens; that is, they can be grown on either wet or dry land, or on land which is only intermittently wet. Informants say that the short-grained cultivars have little tolerance for dryness or high water, and hence are best suited to the mid-level or shallow contour. Since short-grained _mele_ and _serendah_ types are expected to give greater yields than all other varieties during years without extreme dry periods, they are planted on the shallow contour in an effort to maximize yields. Therefore, the general planting pattern is _gangsa puteh_ and _ekar kerbau_ in deep water, _serendah_ and _mele_ in shallow areas, and glutinous rices around the outer edge.

The single rice cultivar most valued by Pesagi farmers is _gangsa puteh_, not because of productivity, which is moderate, but because it is highly dependable over a very wide range of diverse conditions. In 1976, which was the driest year in Pesagi's known history, _gangsa puteh_ gave very poor yields, but nevertheless out produced all other cultivars. Planted in 26 plots, it failed totally on only 1, while yielding 100 _gantang_ or more on each of 7 plots. Overall, _gangsa_ yields per household ranged from 5 to 270, while the mean was only 80 _gantang_ per household. Even with this great variation, _gangsa_ performed better than all other varieties. For example, the _serendah_ group, also a favorite, and a close second in terms of the number of plots planted, failed totally on 4 out of 21 plots. Across households _serendah_ yields ranged from 4 to 100 with a mean of only 35.7 _gantang_ per household. The _mele_ varieties, which informants say in most years yield higher than either _gangsa_ or _serendah_, averaged only 23.6 _gantang_ per household. Thus planting _gangsa_ in deep water and _serendah_ and _mele_ in shallow water is a double strategy designed to maximize yield in both

FIGURE 19 SCHEMES FOR CLASSIFYING RICE

Classification Schemes:	Pesagi Rice Cultivars:*													
	JA	SB	EK	GP	SE	ME	PH	PK	PN	PB	PJ	PD	PG	TL
By general growing conditions:														
(1). Hill rice, dry-land paddy							X	X	X	X				
(2). Lowland, irrigated or flooded	X	X	X	X	X	X	X	X	X	X	X	X	X	X
(3). Floating rice		X	X	X										
Size based on grain length (with husk removed):														
(1). Extra long: over 7 mm	X	X	X	X			X	X	X	X	X	X	X	
(2). Long: 6 to 7 mm					X	X								
(3). Middling: 5 to 5.99 mm														
(4). Short: under 5 mm														X
Shape based on length to breadth:														
(1). Slender: over 3	X	X	X	X			X	X	X	X	X			
(2). Medium: 2.4 to 3					X	X						X	X	
(3). Bold: 2.0 to 2.39														
(4). Round: under 2														X
Type of texture:														
(1). Hard starchy,vitreous fracture	X	X	X	X	X	X								
(2). Soft dextrinous, opaque fracture							X	X	X	X	X	X	X	X
Period of maturity:														
(1). Very early: 110 days or less														
(2). Early: 111 to 140 days														
(3). Late: 150 to 170 days														
(4). Very late: above 180 days	X	X	X	X	X	X	X	X	X	X	X	X	X	X
According to geographical race or subspecies:														
(1). Indica, mostly tropical monsoon	X	X	X	X			X	X	X	X				
(2). Japonica, mostly temperate					X	X								
(3). Javanica, mostly equatorial											X	X	X	

*Same symbols as used in Figures 18 and 20.

normal and drought years.

It is said that _gangsa puteh_ can also withstand high water, and even a week or more of complete submersion. Informants assured me that this would kill or severely damage any other cultivar, with the exception of another deep water cultivar, _ekar kerbau_. These two cultivars have unique stems which are buoyed-up and supported by high water, while other kinds tend to be weighted and broken down. In addition, if the water rises only inches per day, the stems can quickly elongate to keep the tips above water, even though the water depth increases to several times (i.e. 10 or 15 feet) the normal level. This suggests that _gangsa puteh_ and _ekar kerbau_ are similar to what experts have called "floating rice" as grown in flooding areas of Cambodia, the Mekong delta, and elsewhere (Chang _et al_. 1965:23; Grist 1953:33,54; Hanks 1972:35).

It is locally believed that vitreous rices produce better, are healthier and more vigorous, and are more resistant to disease, when grown in close proximity with glutinous rice. This is a major reason why all rice farms contain plots of both types. One local explanation is that Pesagi ancestors believed in the 'rice soul' (_semangat padi_), resident in special _pulut_ rices, which acts as a protective force when given appropriate ritual attention. However, most informants give this "rice magic" explanation little credence, and some object to it as being ridiculous or non-Islamic. A reasonable explanation as to why this belief and accompanying practice continues, is that there are varietal differences which repel or withstand the ravages of adverse growing conditions, and that the local pattern in which cultivars are planted extends some of this benefit over the area of the entire rice field. Glutinous rices are hardy and highly tolerant of weed competion, insects and diseases, and withstand dryness to a much greater extent than all other rices, particularly vitreous rices.

Given the major objective of food production, all the better land is planted to vitreous rice, while poor, generally drier land is planted to glutinous rices. The natural configuration of rice fields is such that poor land is at the outermost edge, and completely surrounds the good land, as is illustrated in Figure 20. Since drought, disease, insects, pigs, monkeys, cattle, some birds, and supernatural forces attack from the outermost edges of the farms, the glutinous rice becomes a protective barrier taking the brunt of the attack.

Rice specialists have learned that a group of long-grained varieties known as "indicas", have the widest geographical distribution of all groups. Indicas are primarily distributed in Ceylon, southern

FIGURE 20

Rice Varieties Planted in 1976 at Pesagi: Paya Bemban and east end of Paya Raja

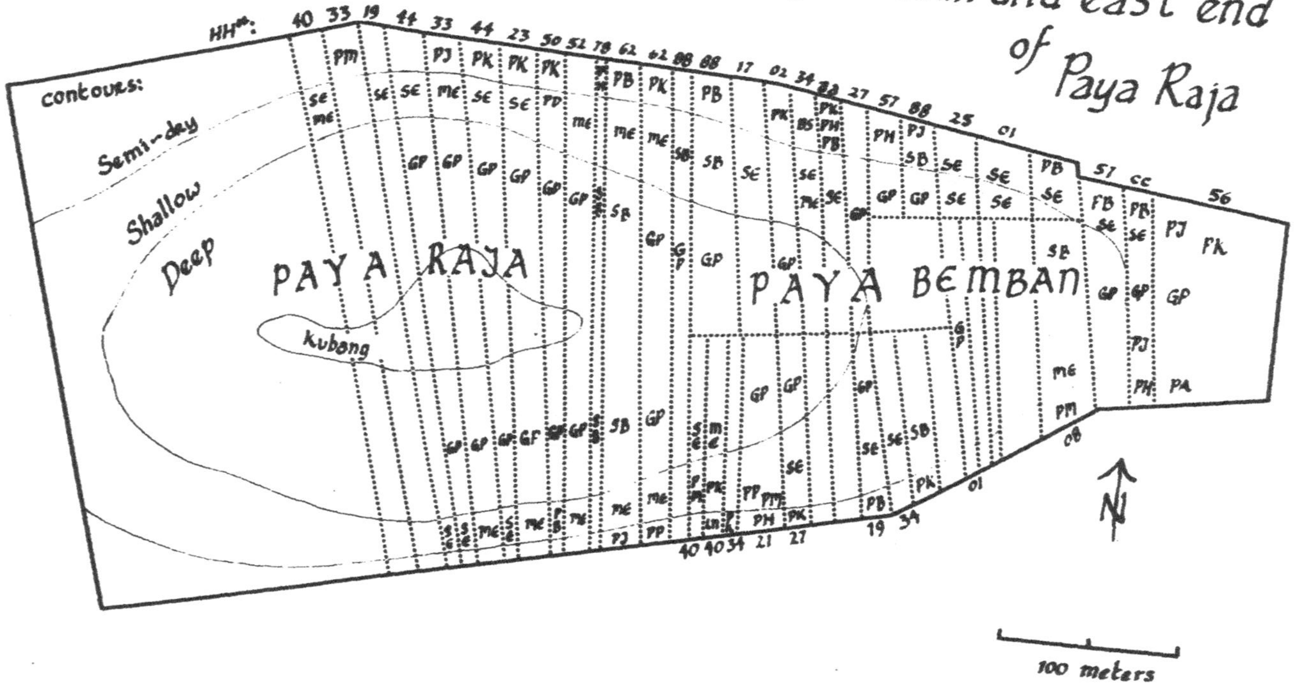

contours:
Semi-dry
Shallow
Deep
PAYA RAJA
Kubang
PAYA BEMBAN
N
100 meters

and central China, India, Java, Pakistan, Philippines, Taiwan, and other tropical areas (Chang 1965 citing Kato *et al.* 1928). The plant breeder Jennings (1966) characterizes the indica group as having

> vast genetic and morphological variability conditioned by diverse environments long prevalent throughout its wide area of cultivation, and by man's limited selection pressures toward the evolution of uniformity in plant type.

Not only are the majority of Pesagi rices indicates, but these exhibit the extreme diversity of colors, textures, and habitat requirements.

The wide range of Pegasi cultivars in the indica group, is a strong indicator that Pesagi farmers practice a form of rice agriculture adapted to a wide range of diverse habitats. By concentrating on indicas, the Pesagi farmer is not required to make laborious or expensive landform modifications (as are essential in *sawah* systems) or to devise means for exact water control. The Pesagi farmer, places little value on improving production methods with expensive inputs (e.g. machinery, chemicals, and "modern" purchased seeds) yet still is able to expand production levels and production area, by using varietials adapted to existing habitats.

The short-grained or "japonica" group, distributed in northern and eastern China, Japan and Korea (Chang 1965 citing Kato *et al.* 1928) is described by Jennings (1966) as contrasting sharply with the indicas because it has

> evolved through environmental pressure and human selection into a relatively homogenous morphological type...

Several informants claimed that the cultivars of this group, namely *mele* and *serendah*, are extremely difficult to distinguish as separate cultivars, because the gross morphological differences are very small. In general, the japonica group, as compared to the indica group is not tolerant of variable water supply. While japonicas usually give higher yields than indicas, they do so only when grown in narrow range of well control-habitats. In Pesagi, only the shallow contour comes close to meeting japonica habitat requirements, approaching in years when rainfall is well distributed the regularity of sawah. Most of these shorter grained types are recent introductions distributed to local farmers by the Agricultural Department following British attempts to improve Pahang yields from about 1920 to 1940 (Birkenshaw 1941; A.O. Report 1929-36).

ADAPTATION PROCESSES

The skillful art of seed selection, and the use of experimental plot trials, are necessary and vital parts of traditional Pahang rice agriculture. It is through these processes that each named cultivar has its distinctive set of locally recognized characters maintained over time, and also that growers constantly improve the cultivars they plant, isolate new ones, and adapt those brought in from elsewhere.

To some extent the large number of cultivars, or even the continued existence of any single one, is dependent upon the behavior of rice growers themselves. Since there is no available supply of rice seed adapted to local environments from outside of Pesagi, the success of next year's crop, and to some extent all future crops, is dependent on the success of each current year's crop and the skill utilized to select and store seed for future planting. From each mature crop, the farmer selects a portion of the best grain for seed. As described by themselves, the choice of which grain to save requires a level of skill which can only be learned through long apprenticeship at the hands of an competent teacher. These teachers, usually experienced older household members, are skilled at recognizing the exact character set which will maintain the pedigree of each specific cultivar. Also, seed selectors are able to occasionally isolate new varieties or improve the yields of old ones.[4]

The closeness of plots to each other, the growing of many cultivars side by side, results in a high frequency of cross-pollination. Close scrutiny of individual plots at harvest time shows that the characters of the seed planted on that plot are exhibited to a greater or lesser degree in at least 85 percent of the plants, while the remaining plants may contrast with the intended cultivar in height, husk color, starch type, number of tillers, disease resistence, and so on. If, harvesters are not extremely skilled, the amount of variation in each succeeding generation can be expected to increase, or the desirable qualities of a particular cultivar can be lost entirely. Informants say that the best seed is owned by households which have had superior seed selectors operating over many generations. Seeds from different households, while bearing the same cultivar name, and exhibiting what are recognized as the same general morphological characteristics, are almost certainly very different in terms of genealogy. For this reason, it was found that informants did not feel that the cultivar names discussed earlier are sufficiently precise, but insist that the source of the seed must be known as well. Thus each harvest of rice seed is distinguished by the name of the household, the harvester, or in some cases the geographical area credited with having developed its distinct pedigree.

Fortunately, each individual household is not entirely dependent upon its own seed selection skill, or the chances of beneficial pollination, or whether or not a planting survives and produces seed -- which it may not if severe flooding, drought, or pest damage occur. One of the things reciprocally exchanged between close kin and friends is high quality seed (and seedlings in years when nurseries do poorly). Theoretically, if each household consistantly uses seed selection methods which improve their own line of seed, and if they abandon poorer lines whenever there is access to better ones, then the accumulative effect, given personal networks covering large areas, is improvement or maintenance of highly adapted and productive cultivars.

Alternatively, a harvester may single out seed from one or more plants, which are perceived to be entirely different from known cultivars, in an attempt to originate an entirely new strain. Examples of attempts to develop new varieties are numerous. _Gangsa puteh ekar musang_, is one particular example, described in Figure 18, as a distinctive black-husked vitreous rice. Farmers would like to grow entire fields of _ekar musang_, not for its color, but because it is more robust, larger grained, and heavier yielding than other _gangsa_ strains. Several households reported having collected and planted _ekar musang_ seed on trail plots. To their disappointment, the results were not very different from plantings of regular _gangsa puteh._ That is, the special seed produced a population in which _ekar musang_ was not significantly increased in proportion. Given the high incidence of cross-pollination, this lack of success is probably because genes producing the desirable characters are recessive, while the frequency of hybridization from adjacent plots continues to be extremely high. Nonetheless, this example demonstrates that growers are skilled at using a technique which undoubtedly with repeated applications over successive generations has successfully isolated new cultivars countless times before.

A quaint explanation for the existance of such a large number of varieties is given by Grist (1953:59-60).

> The perpetuation of such a large number of varieties is probably due to the Asian custom of harvesting ear-by-ear with the implement (_pangani_ in the Philippines, _pisau penuai_ in Malaya) which cuts the ears of paddy singly... The natural result of harvesting the crop ear-by-ear is that

> the attention of the harvester is naturally drawn to any striking variation from the normal. The present writer's experience is that the Asian cultivator is attracted by differences in varieties and is prone to cultivate any paddy which strikes the eye as different, often merely to impress his neighbors. (Has not one seen a similar vanity exhibited by gardeners the world over?) It is, therefore, reasonable to ascribe the persistence of this wide range of varieties as due mainly to the traditional method of harvesting.

From my own perspective it does not seem reasonable to view this kind of "traditional method of harvesting," as having a "natural" relationship to the many varieties of rice. In this case, it is not the implement which accounts for the selection process, but the skill and special knowledge of the person using the implement.[5]

The method of harvest alone does not account for the art of seed selection, or for the cultural and ecological setting which makes possible the continuously adapting agricultural system which results. Clearly, the Pegasi evidence suggests that the large number of varieties is reciprocally related to the range of cultivation skills across individual farmers and the broad range of micro-environments in which rice is planted.

In addition to seed selection, the Pesagi farmer regularly experiments with ways to increase productivity. The local willingness to try new cultivars with expected higher yields is supported by the accounts of informants about experimentation:

> Several gave a government variety, _masuri_, a try, but it had to be abandoned. It gave small yields, and matured so much earlier than other varieties that it was totally eaten by birds.

and,

> Some 30 years ago we used to plant _gangsa tembeling_, a variety brought here from Tembeling (upriver Pahang) by my grandmother. It produced very heavily, but tended to lodge. And it matured so late that it was often damaged by high water. So we stopped planting it.

My field notes recorded several similar seed trials during the 1976 season. One woman planted _pulut dukong_, which she received from a daughter who had married into the Bukit Segumpal Area. Another planted seed obtained from her husband's home at Kampong Awah. This suggests that marriages between Pesagi individuals and outsiders establishes a potential source of new cultivars to be tested locally. It appears that more innovative Pesagi farmers will plant experimental plots of almost any cultivar they can get their hands on, and this includes not only rice but virtually all plants. In my own garden I planted a novel variety of popcorn (Red Strawberry), and at harvest time had numerous requests for seed, some coming from villages in distant areas. Several men proudly showed me their experimental plots of coffee, cacao, black pepper, and oil palm, which they were testing because of government media claims that these will be important cash crops of the future.

Even with well known cultivars the individual household is constantly testing one variety against another, a process of matching varietal performance to small but significant differences in localized habitats. One informant said:

> So, if you have been planting _serendah kuning_ with poor results, you change to some other kind, that's the way to be successful.

The evidence, both in the extensive and diverse range of local cultivars and the attitudes towards new ones, shows that constant experimentation is an integral and important aspect of successful farming as perceived by Pahang Malay farmers themselves.

SUMMARY

The wide range of varieties and habitats helps to make possible the continuation of successful Pesagi rice production, given the uncertainties of weather, disease, and pests from year to year. In a dry year too little water for the shallow contour, or in a wet year prolonged immersion of the deep contour, brings crop loss to each area respectively. In some years rat and wild pig damage is extensive, but confined to the _paya_ edge and shallow water. Pests like army worm (_Leucania_), stem borer (_Diatraea_ sp.), grasshoppers, and birds do scattered damage, but tend to concentrate on only a few areas or a limited number of varieties in any one season. Damage is usually limited to small parts of the total area planted, and farmers can therefore expect a good harvest from one portion or another. The more successful households usually have land in more than one _paya_, and plant two or more varieties to each contour.

Pahang rice farming presents a complex example of diversification in indigenous agriculture. This chapter has focused on rice cultivars, and the next chapter will provide further details about the swamp rice agricultural system. A look at mixed-gardens, the other major type of indigenous farm, shows the same principle to be in effect. That is, many different crops are planted, for each of these crops several varieties are planted. This results in gardens which are adapted to the full range of available habitats, and to irregularities of the market and rapid ecological changes which are inherent to the Pahang river environment. For example, more than 22 kinds of bananas help to supply fruit for local consumption and sale during all seasons of the year. Rather than grow large amounts of any one fruit, which can cause waste because of gluts, the houselot garden may contain as many as 17 genera of fruits, each of which has several species and varieties. There are eleven kinds of bamboo planted for food and construction material. Specimens of 9 kinds of eggplant were collected, and at least that many more were observed. Even chickens, are represented by fifteen or more breeds. Hence, diversification is a major distinguishing feature of Pahang Malay indigenous agriculture.

NOTES

1. An exception is *Terong lembut*, which is not used for human consumption but as bird seed for household pets. *Terong lembut*, has exceptionally small round seeds and appears to be of an entirely different species, probably *Oryza minuta* (described in Burkhill 1966). All other local cultivars are variations of common cultivated rice, *Oryza sativa L.*

2. In a technical sense the *darah belut* and *ekar musang* forms of *gangsa* are not true cultivars, since they are not purposely planted but occur spontaneously in fields of *gangsa puteh*. Local efforts to isolate a strain of *gangsa puteh ekar musang* are discussed under adaptation processes.

3. Imported rice, mostly from Thailand, is very white and highly polished. This *beras Siam* ('Thai rice') has become the standard used for judging the quality of local rice, so that Pesagi farmers often say that their own rice is "just as white as *beras Siam*."

4. Moerman (1968) gives details of a situation which was probably true for earlier Pesagi. In Ban Ping, Thailand, labor and land are so abundant that productivity is not specifically selected for. Rather, farmers plant the local cultivars they most prefer to

eat, and select seed according to "culinary criteria such as aroma, texture, and adhesiveness of the steamed rice."

5. This implement is the same as the *tuai* used in Pahang. It is well suited for harvesting cultivars which ripen unevenly, a characteristic of most local cultivars in which ripening takes a week or more to progress from the central tillers to the outermost ones. Additionally, cultivars with brittle rachis which shatter if harvested with the sickle can be easily harvested with the gentle application of the *tuai*.

5 Making a Living— Swamp Rice Agriculture

There are two kinds of wet rice agriculture practiced in Pahang today. The Pahang Malay, in the central and coastal areas, grow rice in the numerous _paya_ ('swamp lands'). These _paya_ are natural basins watered not by streams, but by direct rainfall. The soft mucky soils are porous, being composed of light river alluvium. The presence of standing water is due to the high water table and frequent replenishment from rain. Farmers working these _paya_ have little or no control over water levels. In contrast, irrigated wet rice agriculture occurs in those areas where terrain is suitable for control of both the supply of water entering fields and the drainage of excess water. These irrigated rice areas are mostly outside the riverine setting of the Pahang Malay, especially near Bentong, Raub, the upper Lipis, and here and there along the Pahang coast. This chapter will describe "swamp rice" agriculture as practiced in Pesagi.

Frequently, those who have described Southeast Asian rice agriculture have taken an extreme macro-analysis approach that simplifies both societies and their ecological settings into two contrasting types: the "wet," intended to refer to 'irrigated' (_sawah_), versus the "dry," which is variously called "slash-and-burn," "hill rice," or "swidden rice." Typically, this simple dichotomy is used to illustrate two possible social-economic or environmental use patterns, as when Geertz (1971) contrasts the "inner" or _sawah_ versus the "outer" or swidden portions of Indonesia, Burling (1965) contrasts the "plains people" using irrigation with the "hill people" using swidden, and Steinburg (1971) refers to Southeast Asian "peasants" as opposed to "upland peoples." In each case, those groups with irrigated rice are seen by the authors as having more complex social organizations, and having more intensive and secure food production systems, than the less intensive swidden societies.

The problem with this conceptualization, is not whether or not it is useful for comparing sawah and swidden systems, but that it is oversimplified, and provides an inadequate basis for understanding the full spectrum of Southeast Asian agricultural practices, ecology, economic development problems, and cultures.

"Wet-dry" models do not take into account local habitat diversity, nor the social and technological aspects associated with the complete spectrum of rice agriculture systems. Many kinds of rice agriculture do not fit easily into a wet-dry dichotomy. For example, three forms of rice agriculture not explicitly included in these models are:

(a) slash-and-burn techniques used for growing rice on swamp lands (in Pahang called simbah);
(b) the use of draught animals on upland dry permanent fields (in Pahang called tenggala, described for the Tausug of Sulu by Kiefer 1975:3); and
(c) wet rice systems using the water naturally available in swamps and marshes (in Pahang padi paya "swamp rice").

The simple dichotomy that Geertz and others have employed can not accommodate the diversity of commonly practiced forms of rice agriculture found in Southeast Asia. This chapter will focus on the Pesagi rice growing system, which could be called "non-irrigated wet rice," "rainfall wet rice," or the "swamp rice" system, and will hopefully provide a broader foundation for the next generation of theoretical models pertaining to Southeast Asian and other rice growing societies.

An examination of the literature on rice agriculture shows that while swamp rice systems are widespread, they have never been given the detailed attention that scholars have devoted to swidden and irrigated rice. This oversight in no way means that swamp agriculture is rare. For example, it has been suggested that swamp rice systems are common throughout the Malay Peninsula, extend to river deltas as far north as the Ganges of India, and are widely scattered throughout most of the Indonesian Archipelago (Takaya et al. 1978:156; Fukui et al. 1978:16).[1] Lucien Hanks (1972:33:36) reports that natural flooding and broadcast sowing occur together in many parts of Thailand, Cambodia, and Malaya, wherever seasonal storms and terrain bring shallow lowland flooding. Over vast areas of Burma's Irrawadi delta rice is grown on natural swamp with little or no control of water supply (Grist 1953:26). In the Chiengrai province of northern Thailand, a wide range of systems coexist to utilize both lands which can be irrigated and the swampy lowlands on river margins (Moerman 1968). In East Malaysia some of the Sadong Bidayu grow rice in "naturally swampy areas" (Geddes 1954:59f.), as do the Malay of the Sarawak River delta (Harrisson 1970; Dunsmore 1968). Earlier reports describe swamp rice in Brunei (Department of Agriculture 1939:55-59), and recent reports show that the practice continues (Allen R. Maxwell, presentation at the Borneo Research Council Meeting, Cincinnati, 1979). There are also recent reports from the Kantu areas of the Kapuas River in South Kalimantan (Noorsyamsi and Hidayat 1974; Dove 1979), and in the coastal swamp lands of Sumatra and Kalimantan inhabited by Banjarese and Buginese (Vayda 1979; Collier 1979). These accounts, while few in number, suggest that lowland non-irrigated wet rice is very widespread both geographically and culturally in Southeast Asia.

There are several reasons for the detailed description of swamp rice agriculture which follows. First, as indicated above, the Pahang Malay practice a highly diversified mixed-garden form of agriculture, and their rice growing system is perhaps the most elaborate and best example. Second, the importance of small-scale production of rice on swamplands is largely overlooked in development projects and agricultural extension programs, at a time when inflation and high costs of living make local food production critically important. Third, Pahang is a natural laboratory for studying and possibly improving a form of rice agriculture adapted to the natural swamplands so widespread throughout Southeast Asia. And fourth, the people of Pesagi take great pride in this distinctive form of agriculture, and hence it is one of the more ethnographically significant characteristics of their culture.

PAHANG RICE PRODUCTION SYSTEMS

The transplanting of rice--planting seed first in nurseries and later moving seedlings into ricefields--has a long history in Pesagi and the other villages near Chenor and Temerloh. _Paya chedong_, which means 'a method for growing rice in the _paya_ that involves transplanting,' is virtually the only rice growing system used throughout the area today, and the local phrase _padi paya_ which means 'any method of growing rice in the _paya_,' commonly refers to the _chedong_ system.[2] Transplanting rice has been practiced for several generations in Pesagi. One of the oldest house clusters in Pesagi, Kampong Chedong, and one of the major swamps, Paya Chedong are named after this agricultural method, and villagers state that _chedong_ has been the predominate rice growing method for "at least five generations".[3] The elaboration of the system, and diversity of adapted rice varieties, might suggest a history of at least several hundred years, but there is little collaborating evidence. Archaeologically it may be possible to correlate agricultural technology with known major foods going back several hundred years, or to associate datable habitation sites with swamp borders. The literature on Pahang agriculture, since 1910, provides clues suggesting that transplanting has been used for many decades, but that its emergence as the only method is a recent occurrence.

The reports on agriculture in Pahang are so brief, and so often inconsistent, that they offer many problems to the student of rice agriculture. While H.W. Jack's lengthy report, "Rice in Malaya," claimed to provide a comprehensive description of rice cultivation in all parts of Malaya, the presentation was generally limited to the highly developed portions of Kedah, Perak, and Province Wellesley, with a lack of detailed information on Pahang (Jack 1923). Others have used a wide range of terms which are poorly defined. R.G. Heath, the Pahang Agricultural Field Officer in 1930 (A.O. Report 1930), reported scattered "small irrigation schemes," the "paya or natural swamp cultivation" in the Temerloh District, "kubang cultivation" in upriver Lipis, and "tenggala cultivation" in the riverine areas. One year later, the field office report of J.M. Howlett (A.O. Report 1931), distinguished only "wet" versus "dry" rice methods, with no description or distribution information. In 1933, a bewildering variety of terms were used by the State Agriculture Officer, J.W. Jolly (A.O. Report 1933), including _padi paya_, _padi kubang_, _padi chedong_, and _padi anak sungei_, all of which were classed as "wet"; and _padi tenggala_, _padi tabor_, _padi ladang_, and _padi tugal_, which were classed as "dry." The major problem in interpreting these reports is that the authors neither defined their terminology nor used consistent classification schemes. If these denotations are given modern Malay interpretations, some refer to natural physical features, some to major human modifications of the local habitat, while still others are horticultural methods. For example:

Natural Features:
paya = swamplands
kubang = deep holes in swamplands
anak sungei = small streams or _alor_

Modified Habitats:
sawah = irrigated pond-like fields
ladang = clearings on dry land

Horticultural Methods:
- _chedong_ = to transplant
- _tabor_ = to broadcast or sow
- _tugal_ = use a pointed stick (often called a dibble) to make seed holes
- _tenggala_ = use a plow

In a 1939 study based on collected reports from British Agricultural Oficers, the Department eliminated much of the confusion by presenting a classification based primarily on a single feature, that being the method used to plant rice seed (Department of Agriculture, 1939). Four "cultivation systems" were distinguished:

> _Tugal_ or _Ladang_ -- Forest or scrub growth on dry land cleared by slash and burn techniques, with seed planted by 'dibble' (_tugal_);
>
> _Paya Simbah_ -- Forest or scrub growth on 'swampland' (_paya_) cleared by slash and burn, with soaked seed broadcast on the untilled soil;
>
> _Paya Tabor_ -- Grass or scrub cover on _paya_ cleared with hand tools such as knives and hoes, with soaked seed 'broadcast' (_tabor_);
>
> _Paya Chedong_ -- Grass or weed growth on _paya_ cleared with hand tools, or sometimes by buffalo trampling. Seelings are raised in dryland nurseries and 'transplanted' (_chedong_) into the _paya_.[4]

There are several indications that, at the time these observations were being made, Pahang agriculture was undergoing rapid change. Also, these so-called "cultivation systems" were, in some cases, short-term transitional stages in the development of upland cash cropping, and not complete or separate cultivation systems within themselves.

The best documented evidence of recent and dramatic changes in Pahang agriculture is for the _tugal_ system, also called _ladang_ or "dry-rice," which is deceptively similar to the upland swidden or "shifting cultivation" systems observed elsewhere, as among the Iban of Sarawak (Freeman 1955) or in scattered parts of Malaya (Jack 1923:114-117). Pahang Malay farmers from the late 20s onward did not use _ladang_ as an on-going production system (i.e. farm to fallow to farm, _ad infinitum_), as in the case of Iban and other swidden system users, but rather removed the forest and established a _ladang_ as a transitory stage for planting long-term, or "permanent" cash crops.

Descriptions of the Pahang Malay _tugal_ system from

early in this century, suggest that rice was only the first of a series of crops grown on clearings destined for long-term cash crops such as coconut, fruits, or rubber.[5] For example, a British Resident's report in 1929 mentions that "hill padi and maize are invariably planted as catch crops [i.e. quick return cash crops] by Malays engaged in opening up new areas for rubber (Cant 1972:117)." A decade later the pattern continued, for it is reported that with _tugal_ cultivation "after 2 or 3 crops are obtained the land is then planted with permanent crops, usually fruits or coconuts (Department of Agriculture 1939:58). The annual reports on Pahang agriculture between 1929 and 1937 (A. O. Report 1929-37) presents statistics and comments suggesting that in general swidden rice was a minor crop except during periods of intensive rubber planting, and that many upland areas suited to swidden cultivation were being planted to cash crops.

The increasing use of dry upland areas for permanent crops forced the increased use of other types of land for rice production: the swamps or marshy lowlands that were unsuited for tree crops. Three of the "cultivation systems" described above were used for wet rice, these being _simbah_, _tabor_, and _chedong_. When viewed as a development sequence, _simbah_ and _tabor_ are possible phases in a transition towards the exclusive use of the _chedong_ system. _Paya simbah_, a kind of swampland swidden, was originally a "long fallow cycle" system requiring a fallow period of nearly 50 years (see Chapter Three). Obviously, the population density would have to remain very low for _simbah_ to be used on a continuing basis. Whenever farmers decreased or eliminated the fallow period, swidden methods were applied to scrub and weed growth rather than forest; and _simbah_ was replaced with the "short fallow cycle" method known as _tabor_. Quite likely, _tabor_ gave low or undependable yields, as young plants had little or no fertilization from ashes, and seeds broadcast on swampy lowlands were vulnerable during short droughts or rains to desiccation or drowning.

It seems reasonable to propose that _chedong_, not a new method but one producing somewhat higher yields with less variability than broadcasting or swidden, became increasingly recognized as the best method for supporting the increasing Pahang population. Harrisson's (1970:559-581) study of the Sarawak Malay describes a similar change from swidden to transplanting in Sarawak during the 1950s, while Hanks (1972) and Moerman (1968) observed the same transformation occurring in Thailand. Hanks (1972:56-7) hypothesized that this was due to a "decreasing variability of output from shifting cultivation to broadcasting to transplanting," which co-occurs with

FIGURE 21
PAHANG RICE YIELDS IN GANTANG/ACRE

YEAR	SAWAH	PAYA (CHEDONG)	SWIDDEN	SOURCE
1915-22	250-350	80-200	104	Jack 1923:143-4
1921		173		Cant 1972:99
1931			119	A. O. Report 1931:52
1932	318	72-201	33-155	A. O. Report 1933:81
1933	310	98-223	52-166	A. O. Report 1933:111
1934	269	77-223	26-122	A. O. Report 1934:82
1933-37			83-191	A. O. Report 1937:48
1933-39			95-168	Grist 1935, 1939, 1940
1939	300-400	190-300		Birkinshaw 1941:28-9
1938			61-72	Barnett 1949:5-6
1956		100-150		Wikkramatileke 1958:24
1965		80-288		Ho 1967:60-1

corresponding increases in yields. The available Pahang data supports this general pattern, as illustrated in Figure 21. In addition to the advantages in yield, Hanks' (1972:167) data also suggest a similar advantage in the amount of labor required per acre. His data shows the difference to be small, however, and there is no Pahang data available for comparison.

The *chedong* system uses some of the techniques common to swidden and irrigated wet-rice agriculture. Rice seeds are first planted in swidden farms (nurseries called *semaian*) which are fertilized through the process of clearing and burning fallow scrub. When about 40 days old, the seedlings are transplanted into *paya* farms, which are located in low-lying basins that catch runoff water and nutrients from surrounding slopes. The nursery usually yields a wide range of products, such as wild produce, firewood, and building materials (made available through clearing), and cultivated fruits and vegetables, along with the rice seedlings. The *paya* farm yields a yearly crop of grain, an abundant supply of fish, and provides grazing land between crops. Together, the two farms are integral parts of a subsistence system which provides not only the major component of the farmer's diet (rice), but also a substantial portion of necessary building materials, medicines, fuel, and other foods.

The process of transplanting seedlings into the *paya* has advantages over broadcast sowing (i.e. the *tabor* system above), and over both dry and swampland swidden. Villagers who have observed or tried several methods report that plants from sown or dibbled seed are frequently choked out by competing weeds, while transplanted rice can outgrow most weed pests. This probably stems from the fact that selected healthy seedlings grown in the rich swidden habitat are being introduced into recently weeded _paya_, and that the larger rice plants form a canopy that helps to prevent subsequent weed growth. Both the roots and the tops of the seedlings are carefully pruned before transplanting and this acts to stimulate growth in the new habitat. Transplanting also promotes tillering (the formation of multiple stems), which results in increased yields (Jack 1923: 173-175). Transplanting makes it possible to control spacing carefully, and to adjust planting densities to localized differences in soil and water depth. For example, one woman informant reports planting bunches of two or three seedlings spaced three per yard in exceptionally fine _paya_, four per yard in ordinary _paya_, and five per yard in poor _paya_. A final advantage of transplanting, is that seedlings can be kept in the nursery for several extra weeks without greatly effecting crop yields (Dore 1960: 40-43), a feature that makes possible the correlation of transplanting to available labor, optimum weather conditions, and _paya_ clearing. Dore's article suggests that this is not true for the cultivars commonly grown elsewhere, but is a unique character of the special varieties which Pahang farming practices have adapted to the _paya_. Varieties planted in _sawah_, for example, are very sensitive to age of seedlings at the time of transplanting.

In terms of the production methods and tools used, Pesagi rice agriculture has changed very little in recent decades. Older informants report little change in techniques during their lifetimes. Many local farmers have observed and even practiced rice growing in places producing greater yields than Pesagi, such as Kuala Lipis, Melaka, Kelantan, and Trengganu, but they explain that most methods originating outside are unsuitable for Pesagi. One man described in great detail the benefits of cooperative work groups in his former home near Lipis, but said that his efforts to organized such groups in Pesagi (his wife's home) were not appreciated, since Pesagi people were accustomed to working only as individuals or as separate households.

Several local farmers have practical knowledge of rice techniques not used locally, such as irrigation, plowing, fertilization, and chemical pest control, but these have not been adopted. There is ample evidence however, that local farmers often conduct small scale experiments to test the feasibility of new production procedures. One man was observed spraying the heavy stand of sedges in his _paya_ with a herbicide, but later reported that dead weeds are more difficult to remove than live ones. Several people mentioned attempts at fertilizing rice nurseries, but concluded that weed growth exceeded that of the rice. As pointed out in the previous chapter, households frequently experiment with new cultivars. There is a considerable interest at present in using small tractors in the _paya_, following an example set in several nearby villages, but many informants feel the cost to be prohibitive, and doubt that mechanical or deep tilling is advantageous. In other villages where machinery is becoming popular, rice growers are often not full-time farmers, but have relatively high cash incomes from wage labor. They grow rice as a pastime or hobby, and they are little concerned with production costs. On the whole, recent innovations are relatively few, the most noticeable being adoption of some improved rice varieties, diesel-powered mills rather than _lesong kaki_ ('foot-powered mortars'), and the widespread use of burlap sacks instead of _tabong_ ('bark bins') for rice storage.

The use of water buffalo to trample weeds into the _paya_ mud or to pull plows is reported as having a long history in Pahang (Wikkramatileke 1958:22; Ho 1967:54). However, there is no good reason to believe that the use of buffalo in rice work was ever an important feature in central Pahang. As mentioned earlier, (see note 1) Takaya and Fukui feel that most _paya_ soils are too soft and too deep to be plowed efficiently or support the weight of buffalo (Takaya _et al_. 1978:156 and Fukui _et al_. 1978:16). Pesagi informants state that they _have_ never used buffalo, as the Pesagi type of _paya_ needs little working, the 'long handled knife' (_tajak_) is easy to use, and because the _tajak_ produces the exact quality of tilth desired. If reports of heavy buffalo loss during floods are accurate (i.e. A.O. Report 1933:65; Winstedt 1927), it is unlikely that plow agriculture based on animal power could be established as a dependable long term technique in the riverine lowlands.

THE CHEDONG SYSTEM AS PRACTICED IN PESAGI

Contemporary rice growing technology in Pesagi and other Pahang riverine areas, is characterized by the use of simple tools. These implements are listed in Figure 22. The extensiveness of this list may appear to suggest an elaborate set of specialized rice production tools, but such is not the case. Only the two items marked (with plus signs) are uniquely used for rice work, while all the others are general purpose tools used in most other household production

FIGURE 22
GLOSSARY OF RICE PRODUCTION IMPLEMENTS

IMPLEMENT	USE
ambong [#1]* 'harvest basket'	collect and transport products
bujang [#3] 'seed basket'	store seed rice or other seed
changkol 'hoe'	loosen soil and cover seed
guni 'burlap sack'	store and carry
ingin padi 'diesel powered mill'	remove husks
kapak 'small axe'	fell trees
lesong kaki 'foot-powered mortar'	remove husks, flour production
niru [#21] 'winnowing basket'	separate product from chaff
parang [#7] 'machete'	clear scrub, trees, and weeds
penggali 'hole digger'	dig post holes
baju paya [#22] 'swamp clothing'	special clothing designed for protection from sun, insects, and especially swamp leeches
tajak [#8] 'long handled knife'	chop weeds and turn soil
tangking [#2] + 'ritual harvest basket'	ritual harvest of first rice
tikar ladang [#17] 'drying mat'	dry rice or other products
tuai [#4] + 'harvesting knife'	reap rice heads

*The numbers in brackets are Field Catalog Numbers for specimens deposited in the Lowie Museum of Anthropology at the University of California, Berkeley, California.

activities.

When informants are asked to describe the sequence of steps used in rice production, they generally emphasize three major and sequential kinds of work: nursery, paya, and harvest work. These are carried out within the time periods discussed in Chapter 3 and illustrated in Figure 10. Each of these kinds of work consists of a set of more-or-less sequential tasks. The chedong system of rice production as practiced at Pesagi is described below, giving first an overview of each major work category, and secondly the work plan for sequencing tasks based on informant statements and field observations. This task sequence plan also lists the major tools and work force members.

Nursery Work

The rice 'nursery' (semaian) plot is cleared using slash and burn techniques, with seed being scattered on the ash and lightly hoed in. Nurseries planted on public nursery reserves are largely mono-crop gardens, for by local convention they can be used only for the brief period necessary to raise seedlings, and therefore crops other than rice are limited to annuals such as maize, pumpkins, and beans. Nurseries on private land, such as houselots or Pesagi Island, are planted not only to rice and annuals, but also with crops productive over several seasons or years and hardy enough to survive careful burning, such as lemon grass, ginger, bananas, sugarcane, pineapple, coconuts, papaya, manioc, spiney palms, pandanus, bamboo, and many kinds of fruit trees. In both cases a fallow of three to six years is required between rice plantings, so that each household must own or have use of an area of potential nursery land equal to several times the area needed per year. My measurements of rice nursery acreage is incomplete, but a reasonable estimate is about one sixth of an acre per year per household. Thus the total acreage required per household is about one acre, including land presently in use plus all fallow plots.

Task Sequence Plan: Nursery Work (Kerja Semaian)

Task	Implements	Workers
A. CLEAR SCRUB (Menebas Belukar):	parang kapak	husband wife

1. clear scrub with knife (menebas),
2. gather rotan, fence wood, firewood, and wild produce from nursery site, and
3. arrange brushwood for even burning.

B. BURN NURSERY (Membakar Semaian): parang, kapak — husband, wife, children

[wait until dry (menunggu kering)]

1. make controlled burns around buildings, fences, and fruit trees or other valuable plants,
2. fire the main garden area, and
3. pile material remaining from first burns and burn again.

C. MAKE NURSERY FENCE (Membuat Pagar Semaian): parang, penggali — husband

1. get needed fencing materials:
   a. get wire [optional if bamboo available] (mengambi dawai),
      i. buy [money saved in advance?], or
      ii. take from other fences or storage.
   b. get posts (mengambi tiang), rattan (mengambi rotan), fence wood (mengambi kayu pagar), and bamboo (mengambi buloh).
2. dig post holes (menggali lubang),
3. set posts (memasang tiang), and
4. tie up fence (mengikat pagar).

D. PLANT NURSERY (Menanam Semaian): parang, changkol, bujang — husband, wife

[wait for rain (menunggu hujan)]

1. perform beginning ceremony (membuat kemula-kemula).

[one day planting taboo (pantang)]

2. broadcast seed [part only] (menabor benih),
3. hoe (menchangkol) seed in,
4. plant vegetables around edges,

[repeat steps 2-4 over a period of two to three weeks following each rain until all seed planted]

5. visit daily to check for animal damage or broken fence.

Paya Work

While seedlings are growing, the farm family spends its time repairing paya fences, and clearing weeds by hand pulling or chopping with parang or tajak. At about 40 days, seedlings are pulled by hand and transplanted to the paya using the fingers to push the roots into the soft mud. During the year of observation the time of transplanting varied considerably from household to household, with all completing the task between mid-May and the end of July. The growing rice is not weeded or tended except to repair fences against cattle and erect traps and devices to limit past damage.

Task Sequence Plan: Paya Work (Kerja Paya)

Task	Implements	Workers
A. MAKE PAYA FENCE (Membuat Pagar Paya)	parang penggali	husband
[same steps as for nursery fence]		
B. CLEAR PAYA (Menerang Paya): 1. clear with tajak (menajak) or clear with knife (menebas), and 2. pile weeds on boundaries between plots.	parang tajak baju paya	husband sons
C. PULL SEEDLINGS (Menchabut Benih):	parang	wife daughters

[when seedlings are past forty days old, wait for rain to flood paya]

1. pull seedlings (menchabut benih),

[continue until 'handful sized bunch' (genggam) is obtained]

2. strike roots against palm to remove soil (mengempas),
3. comb with fingers to remove weeds and dead leaves (menyelara),
4. jostle to make bunch even (mengentok),
5. lay genggam on ground (meletak),

[repeat steps 1 to 5 until sufficient seedlings for transplanting in one day are pulled]

6. gather three to four genggam to make 'large bundles' (unting),
7. tie with bumban (mengikat dengan bumban),
8. prune roots and leaves (mengerat),

[repeat 6 and 7 until all bundled and trimmed]

9. carry bundles to paya (menghantar kepaya),
10. soak in shallow water (merendam kedalam air).

[repeat 9 to 10 until all seedlings set to soak in paya]

D. PLANT SEEDLINGS (Mengubah): baju paya — wife, daughters, husband

[wait for seedlings to soak for two or three days]

1. perform beginning ceremony (membuat kemula-kemula),

[one day planting taboo (pantang)]

2. plant seedlings [shallow varieties like mele and serendah] to shallow contour (menanam paya toho),

[wait for water level to go down, open drainages if necessary]

3. plant seedlings [vitreous] to deep contour (menamam paya dalam),

[restrict drainage]

4. plant seedlings [glutinous and viscous] to semi-dry contour (menaman paya kering).

Harvest Work

During the year of observation harvesting occurred between the second and fourth weeks of December, giving a total paya growth period of 215 to 230 days.

Harvesting is done with a small knife, the tuai, with rice heads being severed one by one and either stacked on top the head, or placed in an ambong basket slung on the back. The newly harvested rice is taken to the houselot, spread on mats to dry, and then 'thrashed by kneading' (mengirek) with the feet. The grain is stored 'in the husk' (padi) and when required for food taken to a diesel powered mill or husked using a 'foot powered mortar' (lesong kaki) to produce 'husked rice' (beras).

Task Sequence Plan: Harvest Work (Kerja Menuai)

Task	Implements	Workers
A. HARVEST (Menuai):	tuai baju paya ambong tangking	wife husband children

[wait for first few heads of rice to ripen]

1. perform ritual harvest of first rice (menuai kemula) using tangking,

[one day harvest taboo (_pantang_)]

2. harvest some immature rice (_menuai padi muda_) for making 'rice flakes' (_empin_),
3. harvest remaining rice as quickly as mature (_menuai padi betol_, lit. 'true rice harvest').

B. DEHYDRATE (_Menjemor_): _tikar_ wife children

[choose hot clear day, or sunny period between rains]

1. spread mats in full sun,
2. distribute rice in thin layer.
3. continuously guard against chickens and cattle.

C. THRASH (_Mengirek_): _tikar_ _niru_ wife husband

1. thrash using feet (_mengirek_),
2. winnow (_mengiru_), and
3. dry again [repeat B].

D. STORE RICE (_Menyimpang Padi_) _guni_ wife husband

store in burlap sacks (_menyimpang dalam guni_), or in wooden bins built inside house.

Work Schedule

Each production activity is carefully coordinated with appropriate weather conditions and seasonal cycles. The new planting cycle begins in February or March, with major effort directed towards planting the nursery with the earliest possible rains, and also transplanting early enough to harvest before flood danger in December and January. In general, no fixed timetable is followed; rather each individual household delays or accelerates production tasks to fit their perception of current and expected weather. During the year of my research work delays were extreme, due to a prolonged shortage of rain early in the season which left the nurseries and shallow _paya_ dry, offsetting the work cycle by about two months (see Figure 23). A work schedule generalized for ideal years was supplied by informants, and is presented below for comparison with the 1976 Pesagi situation.

FIGURE 23
WORK SCHEDULE: COMPARISON OF IDEAL YEAR WITH PESAGI IN 1976

	Ideal Weather		Pesagi 1976	
Activity	Month	Weather	Month	Weather
fence, clear nursery	2-3	mostly dry	3	very dry
plant nursery	2	light rain	4-5	2 rains
fence paya	3	light rain	5	very dry
clear paya, transplant	4-5	heavy rain	6-7	2 rains
	6-9	light rain	8-11	very dry
harvest	10-12	med. rain	12-1	light rain
thrash and store	12-1	heavy rain	1	heavy rain

The 1976 rice season at Pesagi, and throughout central Pahang, was less successful than usual. While droughts are not uncommon, informants say there has never been one comparable to the severity of 1976. One woman, Timah, described conditions as follows:

> We can travel upriver or downriver (i.e. everywhere), and find the 'paya channels' (alor) completely without water. There is absolutely no water at all! How can that be? It's astounding!

These unusual weather conditions precluded rice growing entirely for some households, and brought havoc to the plans of others. A common strategy was to delay final nursery clearing and planting until the weather changed, a strategy which works in most years. But the households that waited for some rain before clearing were unable to plant, because the next rains fell long after the nursery season was past. In general, nurseries did poorly, and grew sickly stunted seedlings heavily infested with weeds. Many households decreased the area of paya planted because of dryness and because of the difficulty of removing weeds from the hardened soil. Transplanting was delayed beyond the optimal 40 days, in some cases beyond 60 days. Much of the transplanted rice died or was badly damaged by dryness. The result was that only one third of the households which started clearing nurseries were able to plant the paya, and only half the planted area produced a suitable crop.

Effective utilization of the paya for rice cultivation requires coordination of planting activities with rainfall. Irrigation from sources outside the immediate paya area is not possible since higher surrounding land is neither very high nor extensive (cf. Birkenshaw 1941:27), and the short Pesagi streams are low and intermittent. Water control consists of attempts to retain whatever rain falls within the _paya_ with dams that can be opened to remove excessive water. Planting begins on the shallow contour first, with water being released as planting proceeds into deeper areas. The danger with this procedure is that subsequent replenishing rains will not occur, and conflicts between individuals wishing to plant large areas and those afraid that precious water is being wasted are common (a case of social problems related to water use is cited in the section following).

Social Organization And Rice Production

In 1976 the distribution of labor among the various rice growing tasks shows that transplanting and harvest require the most labor, but overall labor requirements are quite small. As shown in Figure 24, the average household spent eight days with two people working, and 23 days with only one person working, making a total of 39 worker-days for raising a crop of rice. When compared with worker-days required for producing and marketing rubber, which takes the average household 240 to 280 days per year, the labor requirements for growing rice are very small.[6]

In general, Pesagi residents express a strong aversion to many of the tasks that must be performed when growing rice. Nursery and _paya_ work are said to be extremely dirty and exhausting. Women complain that the thorns of _Mimosa pudica_ scratch their wrists and hands when pulling rice seedlings, and everyone complains about the mosquitoes and leeches in the _paya_. Young women and girls try to avoid rice work because they fear their good looks will be spoiled if they are tanned by the sun. Protective clothing specially designed for rice work consists of a long sleeved shirt and long pants with draw strings at the cuffs, cloth bags to cover the feet, and either the large umbrella shaped hat known as the "Sarawak hat," or a headcloth worn to completely cover the forehead, neck, and sides of the face. Men consider clearing _paya_ using the _tajak_ as a very onerous task. Villagers find rice work to be unpleasant, as work not looked forward to, as simply joyless drudgery. In fact, many commented that if it were not for the high economic value of rice work, they would gladly give it up.

The dislike for rice work has a dampening effect on the expenditure and organization of labor. People work only when they feel they must, and only as much as they feel they have to. For many other tasks, like fishing trips, gathering firewood, or taking crops to market, cooperation is somewhat easy to arrange because

FIGURE 24
RICE WORK: LABOR EXPENDED PER HOUSEHOLD

ACTIVITY	Days	Workers	Worker-Days
NURSERY WORK:			
Clear and Plant	4	2	8
Pull Seedlings	3	1	3
PAYA WORK:			
Clear	6	1	6
Transplant	10	1	10
HARVEST:	4	2	8
OTHER: Thresh, Dry, Store, Mill	4	1	4
	31		39

these activities are enjoyed. The popular image of Southeast Asians working together to plant, weed, and harvest, interspersed with ritual feasting, and so on, is totally alien to the rice fields of Pahang.

1. Division of labor by sex. Rice work can be carried out by either men or women, but for various reasons much of the work falls to women. A few informants defined rice work as strictly women's work, but this view was not widely held, or in any instance exclusively practiced. Clearing and fencing are seen as male tasks, but observation shows that wives often help. Some informants expressed a notion that many kinds of rice tasks are more appropriate for women than for men, because women know more about ancient household rice ritual, women store the rice seed, are the best harvesters, and so on. Observation showed that only in one task, that of pulling rice seedlings from nurseries, women exclusively are the participants. This was explained by saying that only women can properly care for and handle anak padi (lit. 'rice babies'). During interviews in the nurseries women demonstrated, showing meticulous gentleness and care, the specialized steps for pulling and handling rice seedlings which were noted in the nursery task sequence plan. Women carefully instruct their daughters in these skills, and none of the men asked would admit to knowing the specialized details of preparing rice seedlings for planting.

Clearing and fencing are seen as male tasks, but observation shows that wives often help. A major reason males spend less time on rice work than females is because most households must produce cash continuously in order to buy food and satisfy other daily needs. Cash earning is felt to be a basic prerogative and responsibility of household senior males. So, while the senior woman of each household is kept busy almost full-time during the nursery, planting, and harvest periods, men perform rice work only on occasion, or on a part-time basis.

2. Interhousehold work groups. Just as members of each individual household seldom work together, there is also very little cooperative labor between households. Informants say that the use of cooperative labor, such as in 'work bees,' or 'work parties' (bederau), requires greater amounts of labor and higher support cost per worker than working as separate households. Labor requirements are greater because people working together waste time talking and playing. Work quality is poor because people are less likely to do a job well in the fields of others than in their own. Moreover, work parties have to be fed or provided refreshments, which means (because of prestige) a larger quantity and higher quality of food, cigarettes, and drink per worker than would be consumed by a household's own workers when laboring for themselves.

The greatest hindrance to the formation of work parties, or other cooperation between households, is that circumstances often bring about unequal labor exchange. The success of each individual household is often perceived by its members as being so critically related to the coordination of task and weather, that they find it impossible to make good on previous commitments. In a period when weather appears to be steady and favorable, households may readily exchange labor, but with the slightest indication of weather change each deserts the fields of the other and rushes to complete his own work. A general hypothesis is that Pesagi villagers expect the weather to be unpredictable much of the time, and this uncertainty curtails the formation of cooperative parties either entirely or constrains usage to only the most quickly accomplished tasks.

When work parties are formed, members are always intimates. These close kin and friends customarily practice equal reciprocity in labor arrangements and in the exchange of goods. If one household has a poor harvest, other work party members are obligated to share their crops. However, a household with consistently low yields is unlikely to find intimates willing to share production labor. In 1976 only five cooperative labor groups were observed during daily visits to rice fields during the pre-harvest phases (see Figure 25). In all known cases the participants in these parties are close intimates, and the size of

FIGURE 25
RICE WORK: MULTI-HOUSEHOLD WORK PARTIES

PARTY ID#	ACTIVITY	RELATIONSHIP BETWEEN MEMBERS	PARTY SIZE SUM OF HH	DAYS/HH
1	Clear, fence nursery	friends	2	2
2	Clear Paya	father/son-in-law	2	1
3	Clear Paya	friends	2	1
4	Transplant	mother/daughter	2	3
5	Transplant	mother/daughter	2	4

the party is only two households.

It can be concluded that in Pesagi cooperative labor parties are not very important in rice production, since the incidence is very low, work parties are very small, and the amount of labor exchanged is very little.

3. Protecting women. While each household prefers working alone, it cannot grow rice without regard for others. This is particularly the case because much of rice work is done by women. Since women cannot work without protection or guardians nearby (a popular Islamic idea), they must work at the same time when trusted close kin or female friends are present in adjacent fields. Women frequently visit intimates in the late afternoons in order to arrange work schedules which will coincide. Sometimes, when work is critical and no company can be found, a woman's husband or another male relative will take time off from his cash earning activities and help with the rice work.

4. Activity coordination. Some work must be carried out at about the same time across all households planting the same *paya*. Failure to maintain synchrony brings possible crop failure, and also leads to interhousehold conflicts. The most crucial phases needful of coordination are nursery preparation, fence repair, and transpanting. If these phases occur together, the fields of all households mature together, offering the greatest protection against pest damage. According to local experience, a single plot ripening alone is always heavily attacked by birds or rats. If ripening is staggered, each plot in succession may be increasingly damaged, since the continuous supply of food enables pests to multiply. With simultaneous ripening it is said that pest damage is somewhat evenly distributed between plots, and that the crop can be

harvested before pests multiply or are attracted from neighboring districts.

In the case of simultaneous transplanting, there is frequently a problem because individual households fail to complete the planting of each contour at the same time. Only when all households finish planting the shallow contour can water be released so that deep contour planting can begin without creating difficulties. Ho in a study of Pahang Malay near Temerloh states that

> the progressive lowering of water levels has to be agreed upon by all farmers involved, and it is not unusual for more efficient operators to help others with transplanting in order not to hold up operations in the entire _paya_ (Robert Ho 1967:56).

This is possibly a situation in which cooperative labor would be considered worthwhile, especially between intimates. In Pesagi however, such cooperation did not occur, and the lack of synchrony in contour planting led to community conflict and the following incident which I recorded in my field notes:

> During the night of July 5, someone rumored to be from Pesagi Ulu, forcefully broke the lock on the main dam draining Paya Empang. This was said to have occurred because people were not working together on the same schedule, and those who had already completed planting their shallow areas were impatient with the delay in draining water so deep planting could begin.

5. _Fencing of farms_. One of the rice production tasks which is particularly difficult, given the current social organization in Pahang villages, is the fencing of farms as a protection from the ravages of cattle. Lack of fences, or fences in disrepair, is increasingly a serious limiting factor for rice growing. This problem is evident not just in Pesagi, but throughout central Pahang.

Inhabitants state that before the most recent great flood of 1969-70 most villages were very dedicated to rice production. Older informants claim that without exception every single household grew rice every year. This is clearly a statement of the ideal, but not factual, as illness, childbearing, and death, would clearly make it impossible to always generate the necessary labor. Also, it is customary for newly married couples to practice shifting residence between various households of his and her family, and to not get down to serious productive economic activity for several months or until birth of a child. Therefore it is far more likely that not every single could independently manage and provide labor for its own ricefields each and every year. Rather, given the idiom of Pahang Malay, saying that every single household plants every year, is equivalent to saying that most plots of land are planted, since households unable to plant would have their fields farmed by one or another of their close kin.

In the period just before the last flood, communities were dominated by households which were members of extensive intact kin groups, and there were always members looking for rice land to plant. Pressure for land was further intensified by large families. The result was intensive rice land use together with well tended fences. It was not necessary to have a well organized system for enforcing fence maintenance, as a compelling interest in successful rice production brought about the necessary incentive--each household regularly built and repaired its own fence sections. Not repairing one's fence was clearly a very serious offence against the greater community, and was sanctioned with a fine levied by the headman following a public hearing. Apparently, the feeling of responsibility and self interest were sufficient, as informants report that fines were never necessary.

After the flood things changed radically. Some households, particularly wealthy ones with great amounts of land, seem to have permanently moved elsewhere to take advantage of urban based opportunities, though some have been moving back to Pesagi in recent years. At various times, such as during this study in 1976-77, the number of temporarily vacant houses and families coming and going, suggest that up to one-third have adopted a strategy of shifting residence between the village and outside work in towns, factories, or as participants in government agricultural development schemes. The ideal strategy is to return to the village to plant a rice crop, and during rice harvest and when fruit crops mature, and spend the remainder of the year outside earning cash. However, as economic conditions and opportunities fluctuate widely, there are portions of _paya_ without planters in some years. Sooner or later, some sections of fence have no resident caretakers, fall into disrepair, and are breached by animals, bringing about the end of rice production in that particular _paya_. At the time of this study this dilemma was widely discussed in the villages of Pahang, and rice growing families did not see any workable solution. The only alternative was the prohibitively expensive solution already being used around houselots and some other small gardens, that is, each household completely fences the entire perimeter of all its own fields.

THE ABSENCE OF RICE RITUAL IN PESAGI

In many societies rice work has been described as being inseparable from special rituals and magical beliefs about the spiritual essence or "soul" of the rice. Besides serving magical purposes these practices are said to unite the community together at critical times in the agricultural year, and are often occasions for coordinating the timing of production activities, organizing labor groups, and reaching agreements on resource use. The Sarawak Malay have a ritual specialist, the _pawang padi_ ('rice shaman'), who predicts the proper time for planting, and ritually sanctions the location of rice plots for individual households (Harrisson 1970:563). In the Negri Sembilan district of Jelebu some villages have annual ceremonies to placate spirits associated with rice, land, dams, and rivers. According to Swift these rituals have more than a magical purpose, serving also to stimulate cooperation:

> The ceremony is important to the villagers quite apart from the question of whether they believe in it.... At the technical level the need to complete the dam [or other work] before the ceremony can take place is an incentive to get the work finished. Less directly, the ceremony underlines the importance of rice and rice cultivation to the village, removing it from the level of an ordinary activity which anyone can carry out or not as he feels fit. If a man abandons his rubber holding he harms only himself; if he neglects his ricefield, he harms others (Swift 1965:42).

Similar beliefs and practices have also been recorded elsewhere, such as Melaka (Blagden 1897:298) and Kelantan (Hill 1951:60). An important feature of these practices is that they rally together large groups of people, foster cooperation, build social support for rice cultivation, and set a schedule by which individuals or groups initiate and complete each cultivation phase.

In Pesagi I observed only two people performing ceremonies related to rice agriculture, though the sites where ceremonies had been performed were observable in about one third of the nurseries and _paya_ farms. In these ceremonies miniature garden plots, about one yard on a side, with boundaries made of four sticks, oriented with the earth's meridians, are constructed. These ritual plots are located a few feet from one edge of the rice nursery or _paya_ field, and called _kemula-kemula_ ('beginnings') or _tempat bermula_ ('starting places'). In the center of these plots are planted leaves having magical properties: chermaimai to keep away insects, rotan tawar to keep away larvae, and pinang to ripen the crop quickly. After placing these items, the performer recites prayers, and then plants rice to the miniature plot. Traditionally there was a one day work taboo following the starting ceremony, but this is currently disregarded by most planters.

In Pesagi a salient system of rice magic or rice ritual shared by the community, or stimulating cooperation between households is totally lacking. Only a few households performed rituals in 1976. These were always strictly of concern to the performing household alone, and usually required the presence of the performer alone. Today rice rituals are carried out by the oldest residents, while young Malay regard these rituals as pagan, that is non-Islamic, and as serving no practical purpose. Such an attitude serves as a strong deterrent to holding ceremonies, and may also explain why informants declined to give very detailed descriptions of rice ritual and magic. One old man whom I photographed constructing a miniature ritual rice garden, said that except for the prayers (Islamic prayers recited in Arabic) the remainder of the ceremony is just silly nonsense customary to Pesagi people long ago. Observation shows that before beginning any activity, Pesagi residents habitually recite a short prayer, if only the formula Bismillah il-Rahman il-Rahim ('In the name of The Merciful, The Benevolent'), and this Islamic practice has replaced many of the elaborate traditional rituals. Success in rice work appears to be more related to personal and individual faith and practice, or exclusive personal rituals, than to organized group performances.

The social effect of what Pesagi people believe to be proper Islamic behavior, and even traditional rice ritual, contrasts sharply with the example from Negri Sembilan given above. First, there are no group performances that affect scheduling, or promote synchrony. Any synchrony which occurs is not related to ritual, but to the practical efforts of intimates and neighbors to begin critical tasks at about the same time in order to make use of the scarcity of good planting days, available water, and the need to fence adjacent lands against cattle invasion. A second difference, is that Pesagi practice does not raise rice growing to a level different from ordinary activities. In fact, if the modern prayers recited before beginning a task are classified as ritual expressions, they in no way uniquely distinguish rice from any other activity, being identical and equally appropriate expressions habitually uttered before beginning any task.

PRODUCTION RESULTS

Given that 1976 was a year of widespread crop failure, the data on productivity is not a reliable indicator for normal years. However, the data does illustrate several features which informants say are characteristic of rice production in all years. Three of these features will be described in detail. First, rice yields always vary widely across households, years, and plots. Second, a household does not generally produce as much rice as needed for consumption by its members. And third, rice work shows a higher yield in terms of cash value for the amount of labor input, than any other major kind of work.

Productivity, measured as volume yield per acre,[7] varies considerably from year to year, and variability may even increase if abnormal growing conditions occur. In the best years, an acre of land yields 200 plus or minus 75 gantang, a figure based not on Pesagi data -- where yields have been low since the last major flood six years ago -- but on the Pahang Agriculture Department surveys presented earlier in this chapter. The only data available for estimating the yield range for normal years are informant's reports about past harvests. This places the yields for normal years at a mean of about 150 plus or minus 50 gantang/acre. The year of study can be taken as a base year for establishing the lowest possible yields realized during extreme drought conditions. The 1976 data as presented in Figure 26, shows the mean yield to be 76 gantang/acre -- not counting farms with zero yields -- with a range between 7.4 and 248.7.

Almost no one expects to harvest a surplus, that is, more rice than the members of their household can consume. Given the amount of land on which rice is currently grown, and the yields per acre, it is extremely unlikely for a surplus of rice to occur in the Pesagi neighborhood. This can be demonstrated empirically. The total amount of rice required to feed the 600 residents of Pesagi is calculated to be 38,255 gantang per year.[8] If all 96 acres of presently usuable paya (see Chapter Three) were planted, a surplus yield would require over 398 gantang to be harvested per acre. At the normal paya yield level, 150 gantang/acre, Pesagi can only expect to harvest 14,400 gantang, which represents an average deficit of 28,855 gantang per year. Based on these calculations the household of mean size, with ricefields of mean acreage (0.69 acres), can expect in a normal year to harvest only about 37.6 percent (14,400 / 38,253 x 100) of the amount of rice its members require.

While the 1976 data in Figure 26 shows that only six households had yields per acre at the 150 gantang or better level, the data in Figure 27 shows that twelve households managed to exceed the expected 37.6 percent of need. This is largely due to variations in household size and acreage planted per household. For example, Household 34 planted .81 acres (.12 greater than the mean) and had a fair productivity of 154.3 gantang/acre (4.3 above the mean), but due to an exceptionally large adjusted household size of 5.4 persons (1.76 above the mean) harvested only 30.9 percent of their rice needs. While productivity at this level is not far below the expected, it proved to exceed a critical level and was a contributing cause for Household 34 to split into two households. Cash income sources from share-tapping rubber were so low that by mid-season the expectation of a poor rice crop lead the household head to bring a request before the Wakil Ra'ayat for his daughter and son-in-law to be accepted into the development scheme at Jenka Fourteen. This application was approved and reduced the household size to only 3.0.

The most successful household, Household 62, had a surplus harvest of 106.7 percent of their needs, but did so with only a very meager productivity of 79.2 gantang/acre. This exceptional success is due to small household size, 3.0 persons, and a large planting area of 3.03 acres. Access to this much rice land is very unusual, and indeed Household 62 had entirely unusual expectations of its rice, being the only household attempting to produce a surplus to sell. At the normal productivity level Household 63 would have harvested more than twice the amount needed. This is the only household known to have increased its status and wealth in the past few years by concentrating on rice instead of rubber.

In the local view, rice work is seen as being very worthwhile, because of the large volume and cash value of rice which can be won with only a minor expenditure of labor. My data shows that this view is well supported, and is a major reason why the Pahang Malay continue to grow rice for consumption instead of switching entirely to market gardening, rubber, or other cash earning activities. Figure 28 compares the cash value of products from equal labor inputs (i.e. equal numbers of worker-days) in rice work against other work. The "other work" category is largely rubber, but includes all sources of known income such as fish, fruits, coconuts, chickens, and wild produce. The base time period is 39 days, being the mean number of worker-days a household expends annually on rice work. The results show that even with the very poor yield levels of 1976, many households had higher returns for rice work than for other work. Tentatively, the data suggests that households need only produce 30 percent or a little

FIGURE 26 RICE HARVESTED PER HOUSEHOLD: ACREAGE AND YIELD

HH	AREA PLANTED	AMOUNT HARVESTED	PRODUCTIVITY
50	1.19 acre	296 g.	248.7 g./acre
62	3.03	240	79.2
aa	1.28	200	156.2
44	2.82	200	70.9
02	1.02	175	171.6
88	1.34	170	126.8
27	.72	160	222.2
17	1.50	143	95.3
34	.81	125	154.3
52	1.06	124	117.0
04	2.50	121	48.4
63	.94	105	111.7
19	1.22	84	68.8
01	1.33	80	60.1
57	2.18	74	33.9
40	.75	70	93.3
03	2.00	68	34.0
23	1.60	63	39.4
25	.95	63	66.3
56	1.90	56	29.5
cc	.47	42	89.4
11	1.00	35	35.0
78	.44	31	70.5
35	.64	29	45.3
45	.50	20	40.0
33	2.70	20	7.4
08	.84	11	13.1
bb	.27	5	18.5
SUM	37.00	2810	
MEAN	1.32	100.4	75.9
S.D.	.75	74.4	

HH: Household identification number.
g.: Gantang, a local volume unit of measure equal to one British Imperial gallon = about 5.3 lbs. *padi*.

FIGURE 27 RICE HARVESTED PER HOUSEHOLD: COMPARED TO NEED

HH	HH SIZE	NEEDED (A)	AMOUNT OF RICE HARVESTED (B) TOTAL HARVEST	PER CAPITA	MONTHS SUPPLY	PERCENT OF NEED	DIFFERENCE (B - A) g.	$VAL
62	3.0	225 g.	240 g.	80 g.	12.8	106.7 %	+15	+29
27	2.3	173	160	70	11.2	92.5	-13	-25
50	4.7	353	296	63	10.1	83.8	-57	-111
aa	3.2	240	200	63	10.1	83.3	-40	-78
88	3.0	225	170	57	9.1	75.5	-55	-107
03	1.3	98	68	52	8.3	69.3	-30	-58
44	4.1	308	200	49	7.8	64.9	-108	-210
17	3.45	259	143	41	6.6	52.2	-116	-225
02	5.0	375	175	35	5.6	46.7	-200	-388
04	4.0	300	121	30	4.8	40.3	-179	-347
63	3.55	266	105	30	4.8	39.5	-161	-312
cc	1.5	113	42	28	4.5	37.2	-71	-138
52	5.1	383	124	24	3.8	32.4	-259	-502
34	5.4	405	125	23	3.7	30.9	-280	-543
40	3.5	263	70	20	3.2	26.6	-193	-374
19	4.5	338	84	19	3.0	24.8	-254	-493
01	5.0	375	80	16	2.6	21.3	-295	-572
57	4.8	360	74	15	2.4	20.5	-286	-555
23	4.15	311	63	15	2.4	20.2	-248	-481
56	4.2	315	56	13	2.1	17.8	-259	-502
25	5.35	401	63	12	1.9	15.7	-338	-656
78	3.15	236	31	10	1.6	11.9	-205	-398
11	3.9	293	35	9	1.4	11.9	-258	-501
35	3.55	266	29	8	1.3	10.9	-237	-460
08	1.5	113	11	7	1.1	9.7	-102	-198
33	3.7	278	20	5	.8	7.2	-258	-501
bb	1.0	75	5	5	.8	6.7	-70	-136
45	4.05	304	20	5	.8	6.6	-284	-551
SUM		7651	2810				-4841	-9393
MEAN		273.2	100.4	28.7	4.6	36.7 %	-173	-334
S.D.		92.3	74.7		3.5		101	196

HH: Household identification number.
HHSIZE: Household size = persons 20 years of age and over counted as 1, and children calculated as equal to .05 x years of age.
g.: Gantang = local volume unit of measure equal to one British imperial gallon; about 5.3 lbs. padi.
NEEDED: Total rice needed for one year; calculated as HHSIZE x 75 (gantang of padi required per adult per year)(after Correy 1935:529).
$VAL = Local rice value in Malaysian dollars:
1 $M = $US .394 at 1976 value
Vitreous = $ 1.94/g. Glutinous = $ 2.17/g.

FIGURE 28 RICE HARVESTED PER HOUSEHOLD:
VALUE COMPARISON WITH OTHER WORK

HH	TOTAL INCOME OTHER WORK*	CASH VALUE OF 39 LABOR DAYS		PERCENT OF NEED	DIFFERENCE (B - A)
		A OTHER	B RICE	RICE	
50	$ 2295	$ 245	$ 580	83.8 %	$ 335
62	1900	202	475	106.7	273
44	2460	262	392	64.9	130
02	2000	213	341	46.7	128
34	1298	138	246	30.9	108
27	2147	229	310	92.5	81
52	1578	168	241	32.4	73
03	930	99	135	69.3	36
04	2000	213	236	40.3	23
19	1334	142	164	24.8	22
23	1949	204	123	20.2	-81
25	2326	248	123	15.7	-125
35	2160	230	62	10.9	-168
11	2550	272	69	11.9	-203
33	3817	407	40	7.2	-367
SUM	30744	3272	3537		265
MEAN	2049.6	218.1	235.8	36.7%	17.7
S.D.	670.3	71.6	159.0		

* Includes all sources of cash income: rubber, fruit, fish, poultry, wild produce, cakes, and storekeeping.

more of their needed rice in order to make rice competitive with other work. Actually, the relative value of one type of work compared to another is very complex due to variation in household and land size, and in weather, pests, commodity prices, etc. In spite of variation, informants claim that in their experience rice has consistently given a worthwhile return, and therefore is considered in terms of efficiency of labor use to be their most valuable and reliable agricultural crop.

When Robert Ho studied the Temerloh area (1967:57) he recorded rice productivity for 1964 at 1100 lbs/acre, which at 5.3 lbs per gantang, gives a yield of 207.5 gantant/acre. Surprisingly, Ho characterizes a yield of this magnitude as a "dismal performance." This is because Ho's frame of reference is based largely on sawah agriculture, so that he compares Temerloh villages to the average Malayan yield of 440.6 gantang/acre (the best states, such as Kedah, have yields of 550.6, and the worst, such as Trengganu, have 284 gantang/acre). Any Pesagi household would be delighted with the lowest of these yields, and the 1964 Temerloh productivity data when compared with ordinary paya yields would be recognized by any Pahang resident as a truely superb performance. It is totally inappropriate to make the type of comparison Ho has made, without subtracting from sawah productivity the cost of irrigation development and maintenance, fertilizer, tractors or buffalo for plowing, and in many cases hired labor and high land rents.[9] As this chapter has shown, paya agriculture requires little capital and labor, so that it probably more than makes up in efficiency what it lacks in raw productivity.

For many years the Malaysian government rice policy has been to eliminate the need to import from 20 to 30 percent of consumption needs by increasing local production to the self-sufficiency level (Moktar and Mustapha 1975:201). To achieve this goal, great effort and capital have been invested in irrigation projects in Perak, Perlis, Kelantan, and elsewhere. Further expansion of this type is limited by geography, development costs, and the high costs of production. From what I have seen in Pahang, particularly in Pesagi and the several surrounding neighborhoods, a minor amount of relatively inexpensive engineering, and small pumping stations, would likely bring about significant increases in production. It is interesting to calculate what the normal rice production level for Pesagi would be if all its wet lands were utilized. In Chapter Three, it was established that the total paya area is about 450 acres. Of this area, some 100 acres is deeply flooded and requires drainage, and a considerable area is presently too dry because of silt deposits left by floods, so that for technical reasons only 96 acres remain in use. Given drainage where necessary, and limited pumping of water, the entire area could be brought into production. Supposing this greater area were only productive at the normal *paya* rate of 150 *gantang*/acre, then the total yield per year would be 67,500 *gantang*, or a local surplus of 29,245 *gantang* to be sold as a cash crop. This is $56,735.00 worth of surplus rice, at 1976 prices, which would raise the mean income per household more than $400.00. Seen against the income data presented in Figure 28, this is a cash income increase of at least 20 percent. Moreover, it is reasonable to assume that even a small amount of pumping to bring regularity to water levels would decrease the variation in yield levels, and also raise the mean productivity. Even if only minor improvements were made throughout Pahang, these would decrease the amount of rice imported into Malaysia, and definitely help to achieve the government objective of self sufficiency.

NOTES

1. This is the geographical range in which the studies of Takaya, Fukui, and the others, have found the *tajak*, a popular hand tool used in swamps for clearing and tilling. They assert that this tool (also common to Pahang Malay) is adapted for use on swampy or mucky soil in areas where drainage and plowing with water buffalo is not possible. Presumably, tillage must be done by hand because soils are too soft and deep to support the heavy weight of buffalo or to lend itself to efficient plowing.

2. The common local term for 'transplant' is *mengubah*, while *chedong* is reserved for technical or formal usage.

3. My impression is that five generations is a very conservative estimate. The local phrase *dari moyang moyang dulu* (lit. 'before the ancestors'), can either be interpreted as before the lifetime of anyone presently living, or at least four generations before the speaker.

4. Takaya *et. al.* (1978:315-16) describes nearly identical types of rice growing methods for the marginal hills surrounding the Kelantan alluvial plain.

5. However, true swidden systems were common before the cash crop period. Belfield (1902:124-5), for example, reports the *tugal* method being followed by a fallow period.

6. Robert Ho (1967:64-5) calculates labor input per acre in manhours and plots it against yields per acre, indicating that yields respond positively to labor increases until 300 or 350 manhours (36 - 44 worker-days) is reached, but beyond that point further labor inputs bring little or no further yield increases.

7. All of my data refers to rice 'with husks' (_padi_) unless rice 'without husks' (beras) is specifically indicated. See Appendix A for equivalents of local measures.

8. This calculation is based on 140 households with a mean adjusted size of 3.64 consumers, and consuming rice at the rate of 75 _gantang_ per consumer per year.

9. Another very high cost is due to the fact that the government has made it a practice to subsidize rice prices in order to make rice agriculture more attractive. Moreover, since irrigated rice systems are mono-crop systems rather than diversified, in a year with low yields areas such as the Muda irrigation scheme have no other production alternatives, and require substantial government aid. Recent events suggest that large scale irrigated rice projects in Malaysia are economically and politically unstable.

6 Prospects and Conclusions: Diversified Farming— An Adaptive System

PROSPECTS

One thing which struck me on first visiting Pahang was that Pahang Malay villages, except for those close to large towns such as Temerloh, had many outward signs of poverty. Most, such as the village at Tanjong Batu (Paya Pasir) had numerous vacant lots or abandoned houses, and the buildings of residents were frequently patched up with odd pieces of sheet metal and rudely fitting boards. With a few exceptions, kitchen gardens and houselots had run to weeds, and broken fences allowed the ubiquitous grazing of cattle. Throughout central Pahang there was much variation from village to village. Several totally vacant villages were found, near Bukit Lian and Kaula Bera, another close to Durian Hijau, and one on the north side of the Pahang between Pesagi and Chenor. In only a few areas were there outward signs of prosperity and growth, such as well kept homes and motor vehicles, where villages were adjacent to larger towns. In between these extremes lie the vast majority of villages: some apparent signs of poverty, but also signs of wealth; some vacant houses, and some new ones; people moving out of the village, and people moving in. In other words, most of these "between" villages display indications of accommodation and adjustment to rapidly changing conditions.

The adjustments to rapidly changing conditions made by the Pahang people can be elucidated by using the cybernetic model. In such a model an agricultural production system is seen to include within it a "control system", consisting of positive and negative feedback. This feedback of information, energy, and material into the system serves to either bring about increased deviations (positive feedback) or to maintain the system without deviation (negative feedback). The continuation of any system depends on both forms of feedback. In the positive direction are new ideas and new technology, new crop species, significant changes in the natural environment, increased population and productivity, all of which have potential to radically accelerate and increase deviations. Some changes in the positive direction may be beneficial to the biotic community (including its people, of course), but any change which is too rapid or too great in magnitude, can and will cause the disruption of peoples' social institutions (anomie), destroy components of their ecosystem, and disrupt biotic stability and productivity. In extreme cases the people are no longer in control, cannot solve, and do not fully everstand their ever increasing problems. Hopefully, negative feedback steps in long before changes become too rapid or disruptive, and the system is adjusted by redirection and slowing down towards a more balanced (for Pesagi this means trditional) system. Homeostasis, that is, relative constancy, occurs when negative feedback mechanisms are working well.

Positive feedback beyond "certain limits" results in rapid destruction of the system. A cultural system which works well, dynamically adapts within these limits. The government and other agents attempting to institute community assistance programs must clearly know what these "certain limits" are. At what level of intensity do natural phenomena like floods, or practices of cattle herding, bring ecological disruption which exceeds the limits such that neither villagers nor natural systems can with adequate expediency restore a required level of production? Has the indigenous Pahang Malay agricultural production system reached and attempted to exceed the natural carrying capacity of their ecosystem? Or, might it be that all is well with the Pahang Malay system, except that people are n ot satisfied with the standard of living imposed by local natural limits, and therefore would like to expand into new opportunities?

This study raises a serious question for the development scheme model. As seen here, the indigenous farmer is a highly adept individual who manages a family business based on personal experience and access to accumulated community knowledge. But, the farmer who joins a scheme does <u>not</u> really acquire a new home, farmland, and the opportunity to operate a family farming business; rather, he becomes part of a system managed by technicians and experts, where all farm operations are overseen by supervisors, and "farmers" are reduced to little more than share holding laborers. A proud people like the Pahang Malay do not find this very attractive.

Keeping the cybernetic model in mind, let us look at why some villages adapt while others fail. What causes a Pahang Malay village to be vacated? The answer is very complex, and cannot be fully answered here. There are many threats to the traditional riverine Pahang Malay way of life. The further study of these would teach us much about the ways in which the Pahang Malay as a people, and the ecological and economic production systems which they manage, are integrally intermeshed with the larger social, economic, and ecological systems of Malaysia, Southeast Asia, and the biosphere.

Flood Damage

The frequent minor flooding of farmland during the rainy season presents few problems to which Pahang culture is not well attuned. Most changes brought by these floods are desirable. Nutrient laden silt, reduction of pest populations, and watering of fields, function as negative feedbacks to the system, so that fields are improved and production continues at acceptable levels.

The "great floods," which fortunately occur at long intervals only about three times per century, are entirely different in effect, overwhelming the system with intense positive feedback. The radical changes, such as destruction of houses, fruit trees, and domestic animals, and the alterations of soil and land contours, are so dramatic and far reaching that the Pahang people spend years at restoration and readaption of their production system. Most likely, this usually consists of reconstruction of some variant of their original system, but events following recent episodes suggest that many floods trigger a series of major system innovations. For example, the 1926 flood was followed by the intensification of rice production based on transplanting into <u>paya</u> fields, and was the beginning point from which many Pahang Malay report concentrating their efforts on rubber gardening. Likewise, the 1969-79 flood has brought some changes. For the first time many people have looked for employment outside of the village, or sought relocation onto government agricultural development schemes. There is a great difference between the innovations which followed the first flood and those following the second, for in the first instance people were adapting to their flood altered environment, while in the second many were abandoning their home environment and seeking a new one.

By the time of my study five years after the flood, however, it was obvious that a large proportion of those who left Pesagi had only temporarily migrated, or had become successfully involved in both outside and inside economic activities. It was also apparent from official statements that government officials were very displeased with the fact that Pahang Malay flood victims had only temporarily relocated, or periodically shifted residence from government scheme to village, so as to maintain crop production in both locations. Any cultural anthropoligist could have advised against massive relocations into these 'flood housing development projects' (locally called _ruman banjir_), which were located at centralized sites well removed from the river. Peoples throughout the world who utilize intact indigenous agricultural system, have suffered greatly as radical relocation frequently brings on breakdown of normative social institutions and other forms of cultural disruption. It is to be hoped that officials, and others who extol the desirability of "modernization", and the giving up of traditional forms of village life, are not greatly successful.

There is clearly a serious problem here. Are the lands of hundreds of villages along the Pahang River to be abandoned? Can the towns, industries, and development projects absorb 250,000 Pahang Malay? Clearly, a regional strategy for rapid recovery from the effects of great floods must be developed. History shows that great floods pose no great risk to human life, and that the most productive lands of Pahang are near the river. Assistance in restoring indigenous production systems following floods, would be of far greater long term utility than programs designed to move the Pahang Malay elsewhere. Many village areas have land above flood levels where new homes could be built. Farmers need loans of tools and cash, and assistance with the tasks of rebuilding after floods. For the most part they know what their lands can do, but the ability to quickly restore damaged fields is beyond their means. Clearly, great floods are a very great danger to the continuation and further development of a distinctive Pahang Malay way of life, and further study of recovery techniques and the effects of various forms of aid to victims is needed.

"Development" and Related Ecological Damage

The problem of socially and culturally disruptive development projects has just been mentioned above, but here I should like to elaborate on how some of these projects endanger the Pahang Malay way of life through destruction of the natural environment. For the rice farmer domestic cattle are the greatest crop pest in Pahang, and poor farmers who are dependent on rice growing have no way to prevent the wealthy from increasing herd size. The recent increase in the cattle population is largely due to a post flood government program encouraging water buffalo and domestic cattle rearing through the distribution of calves, breeding stock, and the availability of free veterinary services. The few wealthy villages, particularly if they have large rubber holdings or reliable outside income, favor cattle rearing as a prestigious and easy means of expanding capital. It is indeed unfortunate that the government is not cognizant of the fact that this particular development scheme is ecologically unsound, and contributes to rural poverty. In other words, both the rapid "development" of Pahang and the problems with local weather, have coincided to destabilize villages and disrupt local production systems.

A far reaching ecological problem related to development is the proliferation and indiscriminate use of agricultural pesticides, together with the improper disposal of industrial wastes. Pahang Malay in my study, who worked on the Jenka agriculture development schemes, frequently reported headaches, nausea, and skin irritation because of the heavy use of chemicals there. Some reported not feeling well every time they entered the oil palm or rubber fields. Many were regularly exposed on these jobs to dieldrin, 2,4,5-T, mercuric fungicides, and arsenic based herbicides, and their complaints suggest inadequate training, poor supervision, and little or no safety precautions. Some Pesagi villagers were able to obtain these chemicals, from scheme sources, for local use. One villager was hospitalized from eating a duck which was inadvertently contaminated with aresnic, and several people were made ill from eating river fish. Several times during 1975-76 people reported that chemicals and wastes entering the Jenka and Pahang rivers from agricultural development schemes and processing plants produced massive fish kills. There were numerous articles published in the New Straits Times newspaper during 1976, reporting similar pollution problems throughout Malaysia. It is to be wondered how long it will be before people like the Pahang Malay whose way of life makes them directly dependent on their biotic community, will find their rivers and swamps unable to produce fish, their water too poisonous to drink, and the land polluted from the "development" of the region around them.

Economic Competition

Unfortunately for Pesagi and other Pahang villages, government managed economic competition is driving small farmers out of business. Some would see this as part of some sort of necessary "urban transition" in which farm families move to cities to take up factory jobs, rural areas are transformed into highly efficient and productive factory farms, and society and the world is benefitted in the process. However, as the over-abundance of development project rubber has already begun to make Pesagi produced rubber unmarketable, villagers are faced with having no cash crop with which to feed their families. The major alternative presented to such villages is to abandon the village and join a rubber development scheme, such as the one at Jenka Fourteen. This displacement of villagers to further expand production of the same crop elsewhere further exacerbates the problem. Agriculture "development" of this type proceeds at the expense of the village way of life.

If the object of development is to both improve the quality of life and produce revenue earning export products, then the task would be best accomplished through improving existing farms and village production and marketing conditions. That is, real development can only proceed within the limits of the social-economic system of which people are a part, and if the on-going system is abused, we should expect destruciton of the culture and environment. The breakdown of Pahang Malay villages is ample evidence that such destruction is occurring. The "development scheme", when seen in this light, is certainly a form of contra-development for village people, whether or not they remain in the village or work as laborers on scheme lands. Development, if its purpose is to expand or realize the potentialities of the diversity of a nation's people so that their lives will be fuller, better, and greater, should never plunder existing opportunities, but should be dedicated to improving what already works and to providing new opportunities which will elaborate or enlarge the system.

There are certainly several other threats to the continuation of the Pahang Malay way of life, such as the lack of basic business facilitating amenities (roads, electricity, telephones, drinking water, irrigation) in many villages, and an education system which takes children out of their homes and places them in boarding schools in urban areas. This latter threat simply makes it very difficult for parents to teach their children agricultural skills, and complicates adustment to, or appreciation and understanding of, rural lifestyles for the next generation.

CONCLUSIONS

In this book, I have concentrated on how Pahang Malay farmers adapt to ecological changes through their skills and knowledge pertaining to diversified agricultural production. The study of such indigenous systems is not only in the interests of biologists and social scientists, but also has potential in the practical realm of facilitating an increase in culturally and ecologically-adapted programs of agricultural development.

Pahang farmers have brought about a constantly adapting kind of equilibrium within their environment, which is reflected in their patterns of social organization and land tenure, and their resource utilization technology. Ecologist, Eugene P. Odum, argues that flexibility and resilience occur in biological systems because

> the more species present, the greater the adaptation to changing conditions, whether these be short-term or long-term changes in climate or other factors. Or to put it another way, the greater the gene pool the greater the adaptation potential (1963:34).

Informed experts have argued that the persistence of a system increases with technological, biological, geographical, and social diversification (cf., Brookhaven Symposium, 1969). My own study of Pahang farmers suggests that the principal of "increased adaptation potential through diversity" best explains the utility of Pahang production systems and indigenous agricultural knowledge.

The diversity and adaptability is demonstrated in many ways. Each farm is small in area, yet is a mixed-garden with a very wide range of natural and cultivated species. The several kinds of farms (e.g., houselot rubber gardens, and rice nurseries) and their fallow stages, because of close contiguity, serve to increase the biological heterogeneity throughout the entire area of village lands. The farmer's interest in continuing to utilize natural products, promotes the preservation of a highly varied range of semi-natural ecosystems (e.g., some areas of forest, most stream borders and field margins, small river islands and swamps, and the general environs of waterways). Not only do extreme variations in weather not seriously halt production, but economic activities are readily shifted by the highly perceptive farmer so as to capitalize on windfalls associated with changes in rainfall, temperature, river levels, and so on. Finally, Pahang Malay farmers are remarkable for for their achievements in seed selection and experimentation leading to a tremendous range of cultivars, an ability to utilize a great range of available habitats, and steady improvements of productivity. From a long term systems perspective, the overall usage pattern supports the farmer with continuously available produce, maintains species diversity and hence adaptive potential, and ensures the recycling of resources for future use.

It is certain that indigenous Pahang agriculture is a more viable system of production than the externally introduced scheme agriculture. Pahang Malay farmers have so adeptly managed their relationship with the natural environment that they have been able to live continuously in the same villages over many hundreds of years, and to adjust to economic, ecological, and political changes. In contrast, less than two decades of scheme agriculture has proliferated a wide range of social and environmental problems which threaten not only the continuity of these schemes themselves, but also threaten the continuation of the Pahang Malay way of life.

This study suggests that the Pahang Malay, with a population of about two hundred and fifty thousand, are suffering many harmful side-effects from the opening of hundreds of thousands of acres of their home state to "modern" agricultural development schemes. Pahang Malay villages are overpopulated and cannot acquire needed land, because open land is reserved for large-scale scheme farms. Scheme farms compete with village farms: for the same labor force, the same natural resources, the same markets. The local abundance of scheme-produced products in the marketplace severely limits marketing opportunities for villagers. Pollution from scheme agriculture has continuously destroyed wildlife and other natural resources on which villagers, and of course all Malaysians depend. Even with scheme agriculture still young and far from reaching maximum production potential, the income disparity between scheme and village families is already great and increasing.

My work suggests that one major weakness of scheme agriculture stems from the attempts of federal authorities, developers, and agriculturalists to undertake the management and supervision of large numbers of people and large areas of land with the object of reducing the complexity of the natural environment into a single habitat, single crop, continuously high yield regime. As this study has shown, well-adapted agriculture is dependent upon the ability of individual farmers to supervise and manage themselves in accordance with each persons unique needs and available resources. Given that diversity is one key to adaptive agriculture, every individual farmer must be personally familiar with the conditions of his own fields, the attributes of the seed he plants, the availability of labor within his own household and from among his own friends, and be competent in making appropriate decisions when required -- all of which for the most part cannot be fully perceived, understood or effectively managed by anyone who is not a native of the village in question.

One prospect is that development planners will find it increasingly necessary to fully consider the advantages of well-adapted villages and small-scale systems of production. In summarizing a symposium, "Food Energy in Tropical Ecosystems," John W. Bennett argued that it is increasingly likely that many current agribusiness projects in the tropics will fail, and that a fallback to traditional forms of agriculture will occur (presentation American Anthropological Association Meetings, 6 December 1980, Washington, D.C.).

Increasingly, anthropologists, geographers, agricultural economists, political scientists, educators, and others, are arguing that it is an absolute necessity for development planners to take into account the accumulated knowledge and traditional skills and technology of the people among whom they work (cf. Brokensha, Warren, and Werner 1980). The most beneficial development programs, both culturally and in terms of adaptive potential, are those which effectively enhance existing indigenous systems. Towards this end, farmers like those at Pesagi become a major asset, because of their appropriate knowledge, extensive inventory of local adapted cultivars, and long-term experience at the management of ecologically sound production systems.

Pesagi farmers, on their own, have already experienced some success by taking advantage of both the indigenous and the scheme mode of production. Pesagi people perceive these systems as complementary and interrelated, and effectively utilize the advantages of both. The indigenous mode provides security when the less resilient schemes are racked by drought, crop diseases, unhealthy conditions, or low crop prices. The Pesagi farmer seeks to continuously improve village holdings, through innovations brought home from the scheme. This progressive strategy works to improve indigenous agriculture, and greatly reduces the risk of deprivation if schemes fail. Hence, research and development projects should be initiated which seek to improve the already established indigenous agricultural methods. More importantly, and far more fundamental to the modernization of the development process, an atmosphere should be created wherein those special individuals in society which have the cultural abilities and inclination to improve their own production systems are encouraged and supported.

A major thrust of this book has been to suggest that observations in Pahang indicate that Malaysian agricultural development projects are ever increasing the risk of widespread social, economic, and ecological failure, and that indigenous agricultural systems offer a practical and viable alternative. In particular, the prospects are exceedingly good that such well-adapted and highly diversified systems can be enhanced in the direction of increased yields of both export and locally utilized crops, increased economic and ecological stability for the region, and decreased income disparity within the society. If my efforts here serve to improve the understanding of the Pahang Malay, increase the recognition of the importance of indigenous agricultural knowledge, or stimulate further studies of the practical aspects of indigenous technology, then I will regard this as having reached its intended goal.

Appendixes

APPENDIX A FREQUENTLY CITED MEASURES AND CURRENCY

Malaysian		English		Metric
1 gantang (1)	=	1 Br. imperial gallon	=	4.546 liters
	=	1.201 US gallons		
1 pikul	=	133.33 pounds	=	60.479 kilograms
1 kati (2)	=	1.33 pounds	=	.605 kilograms
1 ekar	=	1 acre	=	.4047 hectare

1 Malaysian dollar = .394 US dollars (in 1976)

---

1. A dry volumn measure. When applied to rice: 1 gantang 'with husks' (padi) = about 5.3 pounds (2.4 kilograms) and yields about 3.7 pounds (1.68 kilograms) 'without husks' (beras).

2. Commonly spelled "katty" in English.

APPENDIX B GENERAL HABITAT OF SELECTED PESAGI PLANTS--PAGE 1

Spec.#	Scientific Name*	House lots	Grass land	Paya rice	Scrub land	Rubber farm	Local Name
120	Piper caninum	X					Kadak
	Colocasia esculentum	X		X	X		Keladi
118	Hyptis cf. brevipes	X			X	X	Semoh jantai/Sekutang
	Pandanus sp.	X	X		X		Mengkuang pandan
	Laucaena glauca	X	X		X		Petai jawa
100	Solanum sp.	X	X		X		Terong bulu
103	Solanum sp.	X	X		X		Terong perat
119	Solanum sp.	X	X		X		Terong timbang
	Bambusa/Dendrocalamus sp.	X	X		X	X	Buloh
	Imperata cylindrica	X	X		X	X	Lalang
101	Clidemia hirta	X	X		X	X	Kutu babi
104	Eupatorium odoratum	X	X		X	X	Rumput kepal terbang
162	Mimosa pudica	X	X	X	X	X	Rumput malu
105	Crotalaria cf. striata		X				Rumput gelangan padang
145	Cassia cf. tora		X				Gelingan paya
147	Urena lobata		X				Rumput batang pulut pulut
148	Crotalaria cf. saltiana		X				Bijan
147	Heliotropium indicum		X				Rumput busuk
151	Panicum sp.		X				Rumput chokonit
152	Hyptis ?		X				Rumput batang tumbit
153	Sesamum indicum		X				Rumput batang selaseh
121	Sida sp.		X				Tamsi
150	Cardiospermum halicacabum		X		X		Rumput tebing ayu
102	Asclepias curassavica		X		X		
146	Ageratum cf. conyzoides		X		X		Rumput busuk
157	Setaria cf. geniculata		X	X			Rumput ekor kuching
158	Actrotrema ?		X	X			Bayam rusa
159	Torenia ?		X	X			
154	Panicum ?		X	X			Kumpei
155	Panicum sp.		X	X			Rumput manis
156	Paspalum cf. conjugatum		X	X			Rumput manis
160	Phyllanthus amarus		X	X			

APPENDIX B GENERAL HABITAT OF SELECTED PESAGI PLANTS--PAGE 2

Spec.#	Scientific Name*	House lots	Grass land	Paya rice	Scrub land	Rubber farm	Local Name
161	Hedyotis sp.		X	X			
163	Passiflora foetida		X	X	X		
109	Cyperus sp.			X			Senayan/Umu-umu/Rerabu
106	Cyperus sp.			X			Senayan/Umu-umu
110	Sclera (or Hypolytrum) sp.			X			Senayan/Umu-umu
111	Hypolytrum latifolium			X			Senayan/Umu-umu
108	Eriocaulon cf. sexangulare			X			Senayan/Umu-umu/Rerabu
144	Fimbristylis sp.			X			Mensiang
136	Bulbostylis barbarta			X			Rumput bulu idong
107	Eragrostis sp.			X			Rumput uban
122	Donax cf. arundastrum			X			Bumban air
123	Monochoria hastata			X			Keladi air/K. paya
138	Blyxa cf. malayana			X			Pandan kepau
139	Hydrilla sp.			X			Lumut paya
140	Naias ?			X			Lumut paya alus
141	Leersia ?			X			Rumput kumpai paya
142	Leersia ?			X			Rumput benta paya
143	Ludwigia sp.			X			Maman paya
130	Nepentes gracilis			X	X		Periok kera/Rotan dini
	Licuala sp.			X	X		Palas
129	Lygodium sp.			X	X	X	Reribu
112	Hedyotis philippensis				X		Patah gagok/P. gekgut
164	Rinorea anguifera				X		
114	Tetracera indica				X		Akar mempelas besan
113	Tetracera scandens				X	X	Akar mempelas kesat
115	Aporusa sp.				X	X	Pelangeh/Pelangas
116	Melastoma malabathricum				X	X	Sanudok
135	Allomorphia malaccensis				X	X	Meroyan siti fatimah
117	Ixora cf. javanica				X	X	Petcha priok
124	Saraca declinata				X	X	
126	Psychotria sp.				X	X	Meminyak
127	Osbeckia ?				X	X	Mempoyan

APPENDIX B GENERAL HABITAT OF SELECTED PESAGI PLANTS--PAGE 3

Spec.#	Scientific Name*	House lots	Grass land	Paya rice	Scrub land	Rubber farm	Local Name
133	Clerodendron sp.				X	X	Meluan
134	Champereia cf. marillana				X	X	Chemperai paut
132	Alpinia rafflesiana				X	X	Tepoh kanang
131	kind of fern (Thelypteridaceae)				X	X	Paku
125	Taenitis blechnoides				X	X	Paku
128	Lycopodium cernuum				X	X	Paku merah

---

* The specimens with numbers were identified by Dr. B.G. Stone of the Herbarium of the University of Malaya and a specimem set is deposited with the Herbarium at the University of California, Berkeley, California.

Appendix C Plants Common to Pesagi by Family and Local Use--Page 1

Spec. #	Local Name	Status/Uses (w=wild, c=cultivated p=protected)		Scientific Name
ANNONACEAE				
	Durian belanda	c	fruit, not popular	Annona muricata Linn.
CRUCIFERAE				
	Sawi hijau/S. puteh	c	common vegetable	Brassica chinensis L. var.
LEGUMINOSAE				
162	Rumput malu	w	weed, very widespread	Mimosa pudica L.
105	Rumput gelangan padang	w	weed	Crotalaria cf. striata D.C. [=C. mucronata Desv.]
145	Rumput gelangan paya	w	weed on swamp borders	Cassia cf. tora L.
148	Bijan	w	weed	Crotalaria cf. saltiana
124		w	weed	Saraca declinata
	Kachang panjang	c	vegetable, some sold	Vigna unguiculata (L.) Walp.
	Kachang tanah	c	vegetable, some sold	Arachis hypogaea L.
	Kachang botol	c	vegetable, some sold	Psophocarpus tetragonalobus (L.) DC.
	Kachang parang	c	vegetable, not common	Canavalia ensiformis (L.) DC.
	Kachang bunchis	c	vegetable, some sold	Phaseolus vulgaris L.
	Petai	wp	vegetable, sold	Parkia ?
	Petai jawa	w	rarely eaten	Laucaena glauca
	Jerin	wp	rarely eaten	Pithecellobium jiringa
VIOLACEAE				
164		w	root as med. for rash	Rinorea anguifera
PASSIFLORACEAE				
163		w	common weed	Passiflora foetida L.
CARICACEAE				
	Betek	c	common fruit	Carica papaya L.

Appendix C Plants Common to Pesagi by Family and Local Use--Page 2

Spec. #	Local Name	Status	Uses (w=wild, c=cultivated p=protected)	Scientific Name
CUCURBITACEAE				
	Peria	c	vegetable, some sold	Momordica charantia L.
	Petola	c	vegetable, some sold	Luffa acutangula (L.) Roxb.
	Labu	c	vegetable, some sold	Cucurbita sp.
	Timun	c	vegetable, some sold	Cucurbita sativa L. var.
	Labu air	c	vegetable, some sold	Lagenaria siceraria (Molina) Standl.
DILLENIACEAE				
114	Akar mempelas besan	w	weed in scrub	Tetracera indica Merr.
113	Akar mempelas kesat	w	weed in scrub	Tetracera scandens (L.) Merr.
158	Bayam rusa	w	weed on swamp margins	Actrotrema ?
GUTTIFERAE				
	Manggis	c	fruit, some sold	Garcinia mangostana L.
NEPENTHACEAE				
130	Periok kera/Rotan dini	w	thatch tie, rice contain.	Nepentes gracilis Korthals
PIPERACEAE				
120	Kadak	c	flavoring for rice	Piper caninum Blume
	Sireh china	c	masticant, some sold	Piper betle L. var.
MORACEAE				
	Nanka	c	fruit, some sold	Artocarpus heterpyhllus Lam.
	Chempedak	wpc	fruit, some sold	Artocarpus integer (Thunb.) Merrill
MYRTACEAE				
	Jambu batu	pc	fruit, eat by children	Psidium guajava L.
	Jambu pasir/J. padang/etc.	p	fruit, not popular	Psidium sp.
	Jambu air	wpc	fruit, some sold	Eugenia aquea Burm., E. javanica Lam, or Eugenia malaccensis L.

Appendix C Plants Common to Pesagi by Family and Local Use--Page 3

Spec. #	Local Name	Status	Uses (w=wild, c=cultivated p=protected)	Scientific Name
MELASTOMATACEAE				
101	Kutu babi	w	fruit, eat by children	Clidemia hirta (L.) D. Don
116	Sanudok	w	fruit, eat by children	Melastoma malabathricum L.
127	Mempoyan	w	fence and house supports	Osbeckia ?
135	Meroyan siti fatimah	w	root, birth recovery med.	Allomorphia malaccensis Ridl.
ONAGRACEAE				
143	Maman paya	w	swamp weed	Ludwigia sp.
MALVACEAE				
147	Rumput batang pulut pulut	w	weed, bark as sack tie	Urena lobata L.
121	Tamsi	w	weed in grasslands	Sida sp.
	Kachang bendi	c	vegetable, some sold	Hibiscus exculentus L.
BOMBACACEAE				
	Durian temaga/D. biasa	wcp	fruit, good cash source	Durio zibethinus Murr. var.
OXALIDACEAE				
	Belimbing	c	fruit	Averrhoa carambola L.
	Belimbing batu	pc	fruit	Averrhoa bilimbi L.
RUTACEAE				
	Limau nipis	c	fruit	Citrus sp.
MELIACEAE				
	Langsat	c	fruit, good cash source	Lansium domesticum Jack

Appendix C Plants Common to Pesagi by Family and Local Use--Page 4

Spec. #	Local Name	Status/Uses (w=wild, c=cultivated p=protected)		Scientific Name
EUPHORBIACEAE				
160		w	weed	Phyllanthus amarus Schum. and Thonn.
115	Pelangeh/Pelangas	w	weed	Aporosa sp.
	Ubi kaya	c	tuber and leaf eaten	Manihot esculenta Crantz
	Rambai	wpc	fruit, good cash source	Baccaurea motleyana Hook. form
RHAMNACEAE				
	Bedara	c	fruit, some sold	Zizyphus mauritiana Lamk.
VITACEAE				
	Rambutan telan/R. longkah	c	fruit, good cash source	Nephelium lappaceum L.
SAPINDACEAE				
150	Rumput tebing ayu	w	weed	Cardiospermum halicacabum L.
BALSAMINACEAE				
	Kuini	c	fruit, some sold	Mangifera oderata Griff.
	Mempelam	c	fruit, some sold	Mangifera indica L.
	Pauh	c	fruit, some sold	Mangifera pentandra Hook. form
	Machang	c	fruit, some sold	Mangifera foetida Lour.
OPILIACEAE				
134	Chemperai paut	wp	vegetable, some sold	Champereia cf. marillana
SAPOTACEAE				
	Chiku	c	fruit	Achras zapota L.
ASCLEPIADACEAE				
102		w	weed	Asclepias curassavica L.

APPENDIX C PLANTS COMMON TO PESAGI BY FAMILY AND LOCAL USE--PAGE 5

Spec. #	Local Name	Status	Uses (w=wild, c=cultivated p=protected)	Scientific Name
RUBIACEAE				
112	Patah gagok/Patah gekgut	w	weed	Hedyotis philippensis (Willd. ex Sprengel) [=H. prostrata Bl.]
161		w	weed	Hedyotis sp.
117	Petcha priok	w	weed	Ixora cf. javanica (Bl.) DC.
126	Meminyak	w	weed tree	Psychotria sp.
COMPOSITAE				
104	Rumput kepal terbang	w	weed, very widespread	Eupatorium odoratum L.f.
146	Rumput busuk	w	weed, wound dressing	Ageratum cf. conyzoides L.
CONVOLVULACEAE				
	Ubi keledek	c	tuber and leaf eaten	Ipomoea batatas (L.) Lam.
BORAGINACEAE				
149	Rumput busuk	w	weed	Heliotropium indicum L.
SOLANACEAE				
100	Terong bulu/T. assam	wcp	vegetable, some sold	Solanum sp.
103	Terong perat/T. pipit	wcp	vegetable, some sold	Solanum sp.
119	Terong timbang/T. mangul	wcp	vegetable, some sold	Solanum sp.
	Lada besar	c	vegetable, some sold	Capsicum annuum L. var.
	Lada kechil/L. api	c	veg., good cash source	Capsicum frutescens L. var.
SCROPHULARIACEAE				
159		w	weed	Torenia ?
PEDALIACEAE				
153	Rumput batang selaseh	w	weed	Sesamum indicum L. [=S. oriental L.]
VERBENACEAE				
133	Meluan	w	weed, thatch supports	Clerodendron sp.

APPENDIX C PLANTS COMMON TO PESAGI BY FAMILY AND LOCAL USE--PAGE 6

Spec. #	Local Name	Status/uses (w=wild, c=cultivated p=protected)		Scientific Name
LABIATAE				
118	Semoh jantai/Sekutang	w	weed	Hyptis cf. brevipes Poit.
152	Rumput batang tumbit	w	weed, itch medicine	Hyptis ?
	Api-api	w	weed, ritual uses	Coleus scutellarioides Benth. var.
HYDROCHARITACEAE				
138	Pandan kepau	w	water weed, attracts kepau fish	Blyxa cf. malayana Ridl. [=B. auberti Rich]
139	Lumut paya	w	mossy water weed	Hydrilla sp.
NAJADACEAE				
140	Lumut paya alus	w	mossy water weed	Najas ?
DIOSCOREACEAE				
	Ubi junjung	c	tuber eaten	Dioscorea pentaphylla L.
	Kemeli	c	tuber eaten	Dioscorea pyrifolia Kunth
PONTEDERIACEAE				
137 123	Keladi air/K. paya	w	weed, stems eaten occas.	Monochoria hastata (L.) Solms
ERIOCAULACEAE				
108	Senayan/Umu-umu/Rerabu	w	swamp weed	Eriocaulon cf. sexangulare L.
BROMELIACEAE				
	Nanas biasa/N. morris	c	fruit	Ananas comosus (L.) Merr.
PALMAE				
	Niur/Kelapa	c	fruit, midrib for broom	Cocos nucifera L.
	Pinang	c	masticate, some sold	Areca catechu L.
	Rotan	w	tie material for house construction, fishtraps, fences, etc., fishline	Sp. of Calamus, Plectocomia, Plectocomiopsis, Daemonorops, or Korthalsia

APPENDIX C PLANTS COMMON TO PESAGI BY FAMILY AND LOCAL USE--PAGE 7

Spec. #	Local Name	Status	Uses (w=wild, c=cultivated p=protected)	Scientific Name
ARACEAE				
	Keladi biasa	wc	rarely eaten	Colocasia esculenta (L.) Scott
	Keladi china	wc	rarely eaten	Colocasia sp.
PANDANACEAE				
	Mengkuang pandan	c	weaving mat and basket	Pandanus sp.
	Pandan	c	food flavoring	Pandanus sp.
MUSACEAE				
	Pisang nanka/P. embun/etc.	c	fruit, some sold	Musa paradisiaca var.
ZINGIBERACEAE				
132	Tepoh kanang	w	weed	Alpinia rafflesiana Wall. ex Baker
	Kunyit	c	spice	Curcuma domestica Valeton
	Halia	c	spice	Zingiber officinale
	Lengkuas	c	spice	Alphinia galanga (L.) Swartz./ Languas galanga (L.) Stuntz.
MARANTACEAE				
122	Bumban air	w	weed, tie rice bundles	Donax cf. arundastrum Lour.
CYPERACEAE				
109	Senayan/Umu-umu/Rerabu	w	swamp weed	Cyperus sp.
106	Senayan/Umu-umu	w	swamp weed	Cyperus sp.
110	Senayan/Umu-umu	w	swamp weed	Sclera (or Hypolytrum) sp.
111	Senayan/Umu-umu	w	swamp weed	Hypolytrum latifolium L.C. Rich [=H. nemorum (Vahl) Sprengel]
136	Rumput bulu idong	w	swamp weed	Bulbostylis barbarta (Rottb.) C.B. Clarke
144	Mensiang	w	swamp weed	Fimbristylis sp.

Appendix C Plants Common to Pesagi by Family and Local Use--Page 8

Spec. #	Local Name	Status/Uses (w=wild, c=cultivated p=protected)		Scientific Name
GRAMINEAE				
154	Kumpei	w	weed, grassland & swamp	Panicum ?
155	Rumput manis	w	weed, grassland & swamp	Panicum sp.
151	Rumput chokonit	w	weed on grassland	Panicum sp.
157	Rumput ekor kuching	w	weed, grassland & swamp	Setaria cf. geniculata (Lank.) P. Beauv.
156	Rumput manis	w	weed, grassland & swamp	Paspalum cf. conjugatum Berg.
107	Rumput uban	w	swamp weed	Eragrostis sp.
141	Rumput kumpai paya	w	swamp weed	Leersia ?
142	Rumput benta paya	w	swamp weed	Leersia ?
	Rumput pemunchuk	w	widespread dryland weed	Chrysopogon aciculatus (Retz.) Trin.
	Serai	c	flavoring	Cymbopogon citratus
	Jagong	c	vegetable, some sold	Zea mays L.
	Padi	c	grain, major food	Oryza sativa L. var.
	Buloh	wcp	building material for houses, fences, etc, shoots eaten and sold	Sp. of Bambusa or Dendrocalamus
FERNS				
129	Reribu	w	tying, esp. fishtraps	Lygodium sp.
125	Paku	w	weed	Taenitis blechnoides
128	Paku merah	w	weed	Lycopodium cernuum
131	Paku	w	weed	kind of fern (Thelypteridaceae)
FUNGI				
	Kulat tahun	wp	eaten, good cash source	?
	Kulat keras	w	eaten occasionally	?

Appendix D River Fish Common to Pesagi

Local Name	Use (1)	Max Size kati	Mean Price $/k.	How Caught (2)	Identification (3)
jalawat	sn	40	$10+	c t	Cyprimid *Leptobarbus* sp.
tapah	se	100	$5	t y	Giant Catfish *Wallago* sp.
temelian	se	50	?	t	Cyprimid *Probarbus* sp.
belida	se	15	$2	t y	Featherback *Notopterus* sp.
kaka	se	40	?	t	?
pare	ste	35	?	t	Type of ray
tenggalan	se	3-5	?	t	Murrel *Channa* sp.
kelah	se	"	$2	t x	Cyprimid *Tor* sp.
toman	se	"	?	y	Murrel *Channa* sp.
ubi	sn	"	$4	?	Murrel *Channa* sp.
kerai	se	15	$2	ac p	Cyprimid *Puntius* sp.
patin	se	20	$2.50	acf n x	Catfish *Pangasius* sp.
udang gala	se	1	$3.50	f u	Prawn
tilan	te	"		fj p	Spiny Eel *Mastocembelus* sp.
lampam	e	1		fj p	Cyprimid *Puntius* sp.
kalui	te	3-5		fj p x	Gourami *Osphronemus* sp.
baoung	te	"		fjnpt x	Catfish *Mystus* sp.
lais	e	"		fjnp x	Catfish *Kryptopterus* sp.
buntal	n	3-5		fjnp	Puffer-fish *Tetraodon* sp.
sisah nabi	e	1		fjn	Flatfish *Achiroides* sp.
seluang	e	3-5		fj	Cyprimid *Rasbora* sp.
hara	e	1		p	Cyprimid *Osteochilus* sp.
jaloi	e	"		y	Murrel *Channa* sp.

1 s=some sold, e=eat locally, n=not eaten locally, t=tabu food for some

2 For fishing methods see next page.

3 Tentative identifications using list published by Tweedie (1952).

APPENDIX D RIVER FISH COMMON TO PESAGI (CONT.)

River Fishing Methods:

a=aran [#25]* or rawai [#26]*, fixed line of 50 to 100+ feet, tied to overhanging branches etc., one or many hooks, bait: fruit (Ficus sp.), grasshopper, or earthworm.

c=chodek, fixed line less than 20', tied to pole in mid-stream, one hook, bait: grasshopper. Requires swift water less than 5'.

f=kail, a hand held fishing pole or handline.

j=jala, throw net, for shallow water, and often at night in rain.

n=jaring [#27]*, cross-net or gill-net, used in slow moving water.

p=lukah pengilau, cone-mouthed trap, for shallow flood waters.

t=terubing [#24]*, door-mouthed trap, used where swift water sweeps against bank, best during rain in 2 to 5' rapidly rising or falling muddy water. Sites are scarse with fewer than 10 in Pesagi.

u=lukah udang [#23]*, a cone-mouthed trap, for catching prawns.

x=tajoh [#29]*, fixed fishing pole stuck into mud along shore with hook dangling just at the surface. Bait: (Trichogaster sp.) a fish ikan sepat found in freshwater swamps.

y=tajoh, the same as x, except bait: fruit such as (Fucus sp.).

* Numbers refer to specimens deposited with the Lowie Museum of Anthropology, University of California, Berkeley.

APPENDIX E PAYA FISH COMMON TO PESAGI

Local Name	Use (1)	Max Size kati	Mean Price $/k.	How Caught	Identification (2)
aruan	se	5	$1	bfjns	Murrel *Ophicephalus* sp.
puyu	ste	"	$ .80	bfjn	Climbing Perch *Anabas* sp.
potok	se	1	$ .80	f	?
sembilang	se	"	$1.20	bfjnsx	Catfish *Clarias* sp.
keli	te	3-5		bfjnsx	Catfish *Clarias* sp.
chermin	e	1		bfjnsx	*Ambassis* sp.
kapoh	e	1		bfjn	Perch-like *Pristolepis* sp.
sepat kampodja	e	"		bfjn	*Trichogaster* sp.
sepat biasa	e	"		bfjn	*Trichogaster* sp.
sengiring	e	"		bfjn	Catfish *Mystus* sp.
terubal	e	"		bfjn	Cyprimid *Osteochilus* sp.
bulut	n	3-5		bfjnsx	Swamp-eel *Fluta* sp.

Paya Fishing Methods:

b=*lukah biasa* upright v-mouthed trap, quiet water, 2-3' deep, leaf shade attracts fish.
f=*kail*, hand held fishing pole used in *kubang*, *alor*, and small paya steams.
j=*jala*, throw net, used in *kubang* and *alor*.
n=*jaring* [#27]*, cross-net or gill-net, used in water 2 to 3' deep.
s=*sirkap* [#19]*, basket trap, plunged in water to immobolize fish so it can be caught with hand. Used in *kubang* and *alor* during dry season.
x=*tajoh* [#29]*, fixed fishing pole stuck in mud anywhere in paya, bait: grasshopper.

---

1 s=some sold, e=eat locally, n=not eaten locally, t=tabu food for some
2 Tentative identifications using list published by Tweedie (1952).
* Numbered specimens deposited in Lowie Museum of Anthropology, UC Berkeley.

APPENDIX F COMMON WILD ANIMALS AT PESAGI

Local Name	Use (1)	Status (2)	Habitat (3)	How Caught (4)	Identification (5)
landak	se	c	fsg	s	Porcupine *Hysterix brachyara*
kanchil	se	r	f	s	Mouse-deer *Tragulus javanicus*
munjak	se	vr	f	s	Barking-deer *Muntiacus muntjak*
rusa	se	vr	f	s	Sambhur deer *Cervus unicolor*
tenggiling	e	r	fsg	s	Anteater *Manis jananica*
tikus	np	vc	gh	t	Various rats and mice
tupai	np	vc	gh	s	Tree-shrew *Tupaia glis*
kera	np	c	fsgh	s	Macaque *Macaca fascicularis*
wa-wa	n	c	f	p	White-handed Gibbon *Hylobates lar*
berang-berang	n	r	r	p	Otter *Lutra* or *Amblonyx* sp.
tikus bulan	n	r	fsg	s	Moon-rat *Echinosorex gymnurus* sp.
ular hitam	np	vc	sgh	c	Type of cobra
musang	np	r	fsgh	s	Civit-cat *Paguma larvata*
babi	np	vc	fsgh	s	Pig *Sus* sp.
buaya	np	r	r	t	Crocodile *Crocodilus* sp.
gajah	p	r	f g	p	Elephant *Elephas maximus*
ayam hutan	e	vr	fs	st	Wild Chicken *Gallus gallus*
biawak	np	vc	sgh	c	Monitor Lizard *Varanus* sp.

1 s=some sold, e=eat locally, n=not eaten locally, p=pest
2 vr=very rare, r=rare, c=common, vc=very common
3 f=forest, s=scrub, g=gardens, h=houselots, r=river
4 s=shotgun, t=trapped, c=clubbed or speared, p=protected by government.
5 Tentative identifations based on field observations and Medway (1969).

APPENDIX G   SINGAPORE RUBBER PRICES FOR PAST 70 YEARS

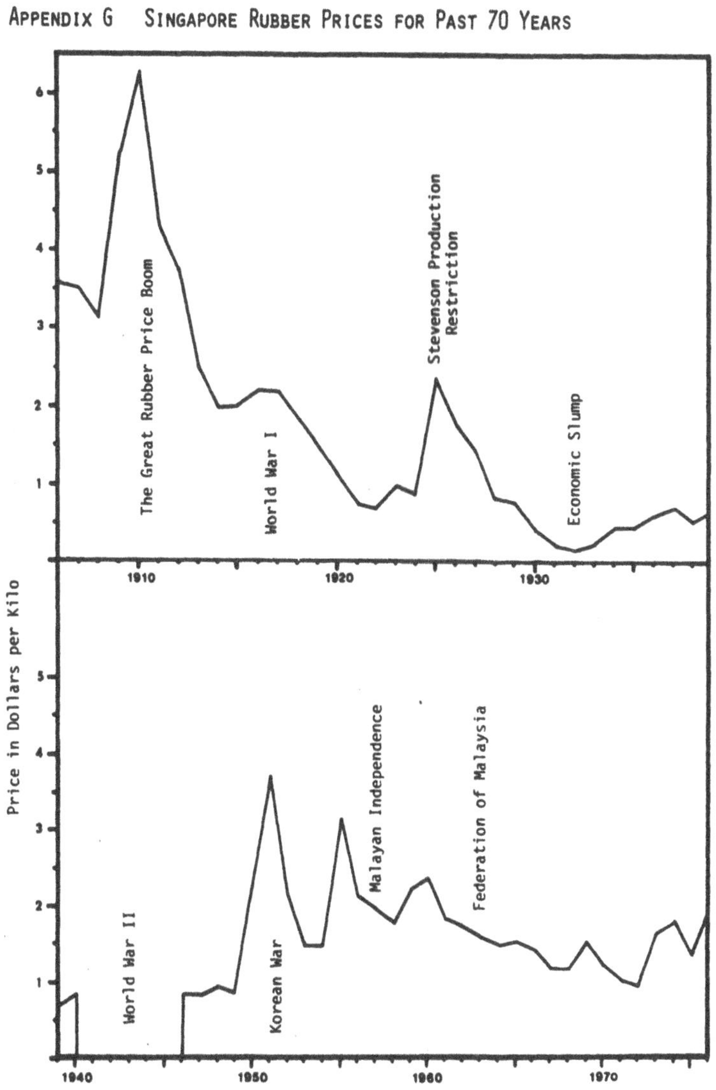

Bibliography

A. O. Report
1929-37 Agricultural Officer Annual Field Reports. Each year authored by the current Agricultural Officer: 1929-30 R.G. Heath; 1931 J.M. Howlett; 1933 J.W. Jolly; 1934-7 H.J. Simpson. Malayan Agricultural Journal.

Abdul Aziz, Ungku
1956 The Causes of Poverty in Malayan Agriculture. Problems of the Malayan Economy: a Series of Radio Talks. Ed. Lim Tay Bon. Donald Moore, Singapore.

Abdullah bin Abdulkadir, Munsi
1893 Kesah Pelajaran Abdoellah bin Abdelkadir Moenshi Dari Singapoera ka Negeri Kalantan. Translated by Brons Middel. E. J. Brill, Lieden.

Alatas, Syed Hussein
1965 Collective Representations and Economic Development. Kajian Ekonomi Malaysia 2(1). Department of Economics, University of Malaya, Kuala Lumpur.

Anderson, James N.
1979 Traditional Home Gardens in Southeast Asia: a Prolegomenon for Second Generation Research. Paper presented at the Fifth International Symposium of Tropical Ecology. Kuala Lumpur, Malaysia, 16-21 April, 1979.

Anderson, James N. and Walter T. Vorster
1976 In Search of Melaka's Hinderlands: Beyond the Entrepot. Paper presented to the Conference on Colonial Port Cities in Asia. University of California at Santa Cruz, 14-16 June, 1976.

Annual Statistical Bulletin
1970-73 Department of Statistics, Malaysia, Kuala Lumpur.

Asian Development Bank
1971 Southeast Asia's Economy in the 1970s. Praeger Publishers, New York, Washington.

Banks, David J.
1976 Islam and Inheritance in Malaya: Culture Conflict or Islamic Revolution? American Ethnologist 2(4):573-586.

Barnett, H. L.
1949 Rice in Malaya, Season 1947-1948. Malayan Agricultural Journal 32(1):4-17.

Bartlett, H. H.
1956 Fire, Primitive Agriculture, and Grazing in the Tropics. Man's Role in Changing the Face of the Earth. Ed. William L. Thomas, Jr. Published for the Wenner-Gren Foundation for Anthropological Research, and the National Science Foundation. The University of Chicago Press, Chicago and London.

Bassett, D. K.
1964 The Historical Background, 1500-1815. In Malaysia: a Survey. Ed. Wang Gangwu. Frederick A. Praeger, New York, Washington, London.

Belfied, H. Conway
1902 Handbook of the Federated Malay States. Edward Stanford, Long Acre, London.

Belshaw, Deryke
1980 Taking Indigenous Technology Seriously: the Case of Inter-Cropping Techniques in East Africa. Indigenous Knowledge Systems and Development, ed. Brokensha et al. University Press of America, Inc., Maryland. Pp. 197-203.

Birkinshaw, F.
1941 A Review of Field Experiments on Padi in Malaya. Malayan Agricultural Journal 29 (1):3-57.

Blagden, C. O.
1879 An Account of the Cultivation of Rice in Malacca. Translation of a text written in 1893 by Inche Muhammad Ja'far, Malay Writer in the Resident Councillor's Office. Journal of the Straits Branch, Royal Asiatic Society 30:285-304.

Bremmer, M. J.
1927 Report of Governor Balthasar Bort on Malacca, 1678. Journal of the Malaysian Branch, Royal Asiatic Society 5(1):9-205.

Brokensha, David W., D. M. Warren, and Oswald Werner
1980 Indigenous Knowledge Systems and Development. University Press of America, Inc., Maryland.

Brown, L. C.
1913 Malay Kampongs in the District of Temerloh, Pahang. Malayan Agricultural Journal.

Brookhaven
1969 Brookhaven Symposia in Biology, No. 22: Diversity and Stability in Ecological Systems. Brookhaven National Laboratory, Biology Department, Upton, New York.

Burkill, I. H.
1966 A Dictionary of the Economic Products of the Malay Peninsula. Ministry of Agriculture and Co-operatives, Kuala Lumpur.

Burling, Robbins
1965 Hill Farms and Padi Fields; Life in Mainland Southeast Asia. Prentice-Hall, Inc., Englewood Cliffs, New Jersey.

Cameron, W.
1885 Extract From a Letter to the Acting Governor, 4 Sept. 1885. Journal of the Straits Branch, Royal Asiatic Society 15:155-57.

Cant, R. G.
1964 Pahang in 1888: the Eve of British Administration. Journal of Tropical Geography 19.

Carey, Iskandar
1976 Orang Asli: the Aboriginal Tribes of Peninsular Malaysia. Oxford University Press, Kuala Lumpur.

Census
1947 Malaya Comprising the Federation of Malaya and the Colony of Singapore; a Report on the Census of Population. Prepared by M. V. Del Tufo. The Government Printer, Federation of Malaya, Kuala Lumpur.

1957 Population Census of the Federation of Malaya. Vol. 1. Department of Statistics, Kuala Lumpur.

Chang, Te-Tzu and Eliseo A. Bardenas
1965 The Morphology and Varietal Characteristics of the Rice Plant. The International Rice Research Institute, Technical Bulletin 4, December 1965. Los Banos, Laguna, The Philippines.

Chomsky, Noam and Edward S. Herman
1979 After the Cataclysm: Postwar Indochina and the Reconstruction of Imperial Ideology. South End Press, Boston.

Clifford, Hugh
1903 In Court and Kampong. London.

Collier, William L.
1979 Social and Economic Aspects of Tidal Swamp Land Development. Symposium on Tidal Swamp Land Development Aspects, Palembang, Indonesia, February 5-10.

Conklin, Harold C.
1967 Some Aspects of Ethnographic Research in Ifugao. Transactions of the New York Academy of Sciences Series 2, 30(1):99-121.

Corry, W.C.S.
1935 The Temerloh Rice Mill. Malayan Agricultural Journal 23(11):528-529.

Courtenay, P. P.
1956 Plantation Agriculture. Frederick A. Praeger, New York and Washington.

Daly, D. D.
1882 Surveys and Explorations in the Native States of the Malayan Peninsula, 1875-1882. London.

Das, K.
1980a Mending Cracks in a Ricebowl. Far Eastern Economic Review 107(8):28-30. (February 22, 1980).

1980b The Communists Zero In. Far Eastern Economic Review 107 (8):28-29. (February 22, 1980).

Department of Agriculture
1939 Padi Planting Methods in Malaya. Compiled by the Economic Branch of the Department of Agriculture, Straits Settlements and Federated Malay States, From Reports of Field Officers. Malaysian Agricultural Journal 27(2).

Dobby, E.H. G.
1951 The North Kedah Plain: a Study in the Environment of Pioneering for Rice Cultivation. Economic Geography 28:287-315.

Dore, J.
1960 The Relation of Flowering and Maturation Period in Some Malayan Rices to Sowing Date and Latitude. Malayan Agriculture Journal 43(1):40.

Dove, Michael R.
1979 The Swamp Rice Swiddens of the Kantu' of West-Kalimantan. Prepared for the Fifth International Symposium of Tropical Ecology. Kuala Lumpur, 16-21 April, 1979.

Drabble, J. H.
1973 Rubber in Malaya, 1876-1922; the Genesis of the Industry. Oxford University Press, Kuala Lumpur, New York.

Dunn, Frederick L.
1964 Excavations at Gua Kechil, Pahang. Journal of the Malaysian Branch, Royal Asiatic Society 37(2):87-124.

1966 Radiocarbon Dating of the Malayan Neolithic. Proceedings of the Prehistoric Society 32:352-353.

1970 Cultural Evolution in the Late Pleistocene and Holocene of Southeast Asia. American Anthropologist 72:1041-1054.

Dunsmore, J. R.
1968 Experiments on Improved Wet Rice Cultivation in Sarawak Malaysia. International Rice Commission Newsletter 17(4):1-12.

Firth, Raymond
1966 Malay Fishermen: Their Peasant Economy. Routledge and Kegan Paul Ltd., London. Other editions: 1946, 1968, and 1971.

Fraser, Thomas M.
1966 Fishermen of South Thailand: the Malay Villagers. Holt, Rinehart and Winston, New York.

Freeman, J. D.
1955 Iban Agriculture. Her Majesty's Stationery Office, London.

1958 The Family System of the Iban of Borneo. Cambridge Papers in Social Antrhopology 1:15-52.

Fukui, Hayao and Yoshikazu Takaya
1978 Some Ecological Observations on Rice-growing in Malaysia. South East Asian Studies 16(2):189-97.

Geddes, W. R.
1954 The Land Dayaks of Sarawak. Her Majesty's Stationery Office, London.

Geertz, Clifford
1971 Agricultural Involution: the Process of Ecological Change in Indonesia. First printed 1963. University of California Press, Berkeley and Los Angeles, California.

Geoghegan, William H.
1969 Decision-Making and Residence on Tagtabon Island. Working Paper No. 17, Language-Behavior Research Laboratory. University of California, Berkeley.

1973 Natural Information Processing Rules: Formal Theory and Applications to Ethnography. Monographs of the Language-Behavior Research Laboratory, No. 3. University of California, Berkeley.

Gill, Ranjit
1977 New Lease of Life for Rubber. Far Eastern Economic Review (December 9, 1977).

Ginsburg, Norton S. and Roberts
1980 Cognitive Strategies and Adoption Decisions: A Case Study of Nonadoption of an Agronomic Recommendation. Indigenous Knowledge Systems and Development, ed. Brokensha et al. University Press of America, Inc., Maryland. Pp. 9-28.

Goodenough, Ward E.
1956 Residence Rules. Southwestern Journal of Anthropology 12(1):22-37.

Gorman, Chet F.
1969 Haobinhian: A Pebble-tool Complex with Early Plant Associations in Southeast Asia. Science 163:671-673.

Grist, D.H.
1935 Rice in Malay in 1934. Malayan Agricultural Journal 23(1):4-9.

1939 Rice in Malaya in 1938. Malayan Agricultural Journal 27(3):99-105.

1940 Rice in Malaya in 1939. Malayan Agricultural Journal 28(4):164-170.

1953 Rice. Longmans, Green and Co., London, New York, Toronto.

Gullick, J. M.
1958 Indigenous Political Systems of Western Malaya. Monographs of Social Anthropology, No. 17. London School of Economics, London.

1963 Malaya. Ernest Benn Limited, London. Revised in 1969 and retitled Malaysia.

Hall, D.G.E.
1968 A History of Southeast Asia. London.

Hanks, Lucien M.
1972 Rice and Man: Agricultural Ecology in Southeast Asia. Aldine-Atherton, Inc., Chicago and New York.

Harrisson, Tom
1970 The Malays of South-west Sarawak Before Malaysia; a Socio-ecological Survey. Michigan State University Press, East Lansing.

Henderson, H. R.
1974 Malayan Wild Flowers. Malayan Nature Society, Kuala Lumpur. First printed in 1954.

Hill, A. H.
1951 Kelantan Padi Planting. Journal of the Malayan Branch, Royal Asiatic Society 24(1):5676.

Ho Kwon Ping
1976a Malaysia: Using Up the Wastes. _Far Eastern Economic Review_ 93(35):28-42. (July 9, 1976).

1976b Rubber Agreement is Ready for Launching. _Far Eastern Economic Review_ 93(35):74-75.(August 27, 1976)

1976c Palm Oil: The Wonder Crop Will Produce Magic Again. _Far Eastern Economic Review_ 93(35):44-46.(August 27, 1976).

1980a Victims of the Green Revolution. _Far Eastern Economic Review_ 108(25):103-106.(June 13, 1980)

Ho, Robert
1967 _Farmers of Central Malaya_. Research School of Pacific Studies. Department of Geographcy Publication G/4 (1967). Australian National University, Camberra.

Hodder, B. W.
1959 _Man in Malaya_. University of London Press Ltd., Warwick Square, London.

Hsuan Keng
1969 _Orders and Families of Malayan Seed Plants_. University of Malay Press, Kuala Lumpur.

Husin Ali, Said
1964 _Social Stratification in Kampong Bagan: A Study of Class, Status, Conflict and Mobility in a Rural Malay Community_. Monograpoh 1, Malaysian Branch of the Royal Asiatic Society.

1968 Patterns of Rural Leadership in Malaya. _Journal of the Malaysian Branch, Royal Asiatic Society_ 41(1):95-145

Igbozurike, M.U.
1971 Ecological Balance in Tropical Agriculture. _Geographical Review_ 61:518-529.

Jack, H.W.
1923 Rice in Malaya. _Malayan Agricultural Journal_ 5:7-9.

Jackson, James C.
1968 _Planters and Speculators; Chinese and European Agricultural Enterprise in Malaya, 1786-1921_. University of Malaya Press, Kuala Lumpur.

Jeffries, Sir Charles
1956 The Colonial Office. George Allen and Unwin Ltd., London.

Jennings, Peter R.
1966 Evaluation of Partial Sterility in Indica x Japonica Rice Hybrids. The International Rice Research Institute, Technical Bulletin 5 (April, 1966). Los Banos, Laguna, The Philippines.

Keesing, Roger M.
1967 Statistical Models and Decision Models of Social Structure: A Kwaio Case. Ethnology 6(1):1-16.

Kennedy, J.
1962 A History of Malay; A.D. 1400-1959. St. Martin's Press, New York.

Kiefer, Thomas M.
1975 Tausug. Ethnic Groups of Insular Southeast Asia; Volume 2: Philippines and Formosa. Frank M. Lebar, Editor and Compiler. Human Relations Area Files Press, New Haven.

Kuchiba, Masuo
1978 Socio-Economic Changes in a Malay Padi-Growing Community (Padang Lalang) in Kedah. South East Asian Studies 16(2):22-39.

Lamb, Alastair
1964 Early History. Malaysia; A Survey. Wang Gangwu, ed. Frederick A. Praeger, New York, Washington, London.

Linehan, W.
1936 A History of Pahang. Journal of the Malayan Branch, Royal Asiatic Society 14(2).

McGee, T.G.
1964 Population: a Preliminary Analysis. Malaysia; A Survey. Wang Gangwu, ed. Frederick A. Praeger, New York, Washington, London.

McHale, T.R.
1967 Rubber and the Malaysian Economy. University Handbook Series-3. M.P.H. Publications Sendirian Berhad, Singapore.

Maeda, Narifumi
1978 The Malay Family as a Social Circle. South East Asian Studies 16(2):40-69.

Malaysian Government
1971 Second Malaysia Plan 1971-1975. Printed at the Government Press by Mohd. Daud bin Abdul Rahman, Government Printer, Kuala Lumpur.

Map
1973 Malaysia Barat: Pahang. Scale 1:253,440. Printed by the Directorate of National Mapping, Malaysia.

Medway, Lord
1969 The Wild Mammals of Malaya and Offshore Islands Including Singapore. Oxford University Press, Kuala Lumpur, Singapore, and London.

Meilink-Roelofsz, M.A.P.
1962 Malacca at the End of the 15th Century: Structure of Trade. In Asian Trade and European Influence. The Hague.

Mikluho-Maclay, M. de
1878 Ethnological Excursions in the Malay Peninsula, Nov., 1874 to Oct., 1875. Journal of the Straits Branch, Royal Asiatic Society 2:205-221.

Miller, George A., Eugene Galanter, and Karl H. Pribram
1960 Plans and the Structure of Behavior. Holt, Rinehart and Winston, Inc.

Millikan, Max F. and David Hapgood
1967 No Easy Harvest: the Dilemma of Agriculture in Underdeveloped Countries. A report on the Conference on Productivity and Innovation in Agriculture in the Underdeveloped Countries held at Endicott House in Dedham, Massachusetts, from June 29 to August 7, 1964. Little, Brown, and Company, Boston.

Mills, J.V.
1930 Eredia's Description of Malacca, Meridional India, and Cathay. Journal of the Malaysian Branch, Royal Asiatic Society 8(1):1-288.

Mission
1954 Mission of Enquiry on the Industry of Malaya. Government Press by G.A. Smith, Kuala Lumpur.

Moerman, Michael
1968 Agricultural Change and Peasant Choice in a Thai Village. University of California Press, Berkeley and Los Angeles.

Moktar bin Tamin and N. Hashim Mustapha
1975 Kelantan, West Malaysia. Changes in Rice Farming in Selected Areas of Asia. International Rice Research Institute, Los Banos, Philippines.

NST (New Straits Times, Malaysia)
August 24, 1976 Economic Talks Told: Felda Schemes 'a Ripple in Ocean.'

October 2, 1976 Timely Controls.

October 31, 1976 Laws to Control Agro Pollution.

November 1, 1976 Polluted Rivers

Noorsyamsi and Oemer Hidayat
1974 Tidal Swamp Rice Culture in South Kalimantan. Proceedings of the International Seminar on Deep-water Rice, August 21-26, 1974. Bangladesh Rice Research Institute, Joydebpur, Dacca.

Ooi Jin-Bee
1963 Land, People and Economy in Malaya. London.

Ouchi, Tsutomu et al.
1979 Farmer and Village in West Malaysia. University of Tokyo, Faculty of Economics, Tokyo.

Peacock, B.A.U., and F. L. Dunn
1958 Studies in the Prehistory of Pahang. Paper presented at the Fourth International Conference on Asian History, University of Malaya, Kuala Lumpur, 5-10 August 1968.

Peyman, Hugh
1980a Showpiece. Far Eastern Economic Review 107(8):30. (February 22, 1980).

1980b Of Rice and Anxious Men. Far Eastern Economic Review 107(8):41-44. (February 22, 1980).

1980c Bringing Smiles to the Villages. Far Eastern Economic Review 107(10):85. (March 7, 1980).

1980d A Message from the Farmers. Far Eastern Economic Review 108(16):61-62. (April 11, 1980).

1980e Hard Times on the Plantation. Far Eastern Economic Review 108(20):58-60. (May 9, 1980).

Provencher, Ronald
1971 Two Malay Worlds: Interaction in Urban and Rural Settings. Research Monograph Series. Center for South and Southeast Asian Studies, Berkeley, California.

Purchall, J.T.
1971 Rice Economy: a Case Study of Four Villages in West Malaysia. University of Malaya Press, Kuala Lumpur.

Purcell, Victor
1965 Malaysia. Thames and Hudson, London.

Purseglove, J.W.
1972 Tropical Crops: Monocotyledons. Wiley and Sons, New York.

1974 Tropical Crops: Dicotyledons. The English Language Book Society and Longman, London.

Quinn, Naomi
1975 Decision Models of Social Structure. American Ethnologist 2:19-45.

R.N.
1975 Pest Control Comes of Age in Malaysia. The Economic Bulletin 2(7):7-12. Academic Publishers Snd. Bhd. Room 1703 Bangungan Fitzpatric, 86 Jalan Raja Chulan, Kuala Lumpur.

Randall, Robert A.
1977 Change and Variation in Samal Fishing: Making Plans to 'Make a Living' in the Southern Philippines. Doctoral Dissertation (Anthropology), University of California, Berkeley.

Rawson, R.R.
1963 The Monsoon Lands of Asia. Aldine, Chicago.

Ridley, H.N.
1894 Account of a Trip up the Pahang, Tembeling, and Tahan Rivers, and an Attempt to Reach Gunong Tahan. Journal of the Straits Branch, Royal Asiatic Society 25:33-56.

Roff, William R.
1973 Islam as an Agent of Modernization: An Episode in Kelantan History. Modernization in Southeast Asia. Hans-Dieter Evers ed. Oxford University Press, Kuala Lumpur.

Ronquillo, Bernardino
1971 The Banana Boom. Far Eastern Economic Review 73(39):53-4. (September 25, 1971).

Schlegel, Gustave
1899 Geographical Notes: VIII. Pa-hoang, Pang-k'ang, Pang-hang: Pahang or Panggang. T'oung Pao Archives: Pour Servir a l'Etude de l'Histoire, des Langues, de la Geographie et de l'Ethnographie de l'Asie Orientale (Chine, Japon, Coree, Indo-Chine, Asie Centrale et Malaisie). E. J. Brill, Lieden.

Skeat, Walter William
1900 Malay Magic; an Introduction to the Folklore and Popular Religion of the Malay Peninsular. Barnes and Noble, Inc., New York.

Skinner, A. M.
1878 Geography of the Malay Peninsula. Journal of the Straits Branch, Royal Asiatic Society 1.

Solheim, W. G., II
1968 Early Bronze in Northeastern Thailand. Current Anthropology 9:59-62.

Steinberg, David Joel
1971 In Search of Southeast Asia: a Modern History. Praeger Publishers, New York.

Sultan Ahmad
1899 Letter from the Sultan Ahmad Mu'azam Shah to the Chiefs at Jelai and Lipis. Published in History of Pahang by W. Linehan (1936:223-4).

Sutlive, Vinson
1978 The Iban of Sarawak. AHM Press, North Arlington Heights, Illinois.

Swettenham, Frank A.
1885 Journal Kept During a Journey Across the Malay Peninsula. Journal of the Straits Branch, Royal Asiatic Society 15:1-37.

Swift, M. G.
1965 Malay Peasant Society in Jelebu. London School of Economics Monographs on Social Anthropology. No. 29. The Athlone Press, University of London.

Takaya, Yoshikazu, Hayao Fukui, and Isamu Yamada
1978 Ecology of Traditional Padi Farming in West Malaysia. South East Asian Studies 16 (2): 309-334.

Thio, E.
1957 The Extension of British Control to Pahang. Journal of the Malaysian Branch, Royal Asiatic Society 30 (1).

Tweedie, M.W.F.
1952 Malay Names of Fresh-water Fishes. Journal of the Malaysian Branch, Royal Asiatic Society 25 (1):62-67.

1953 The Stone Age in Malaya. Journal of the Malaysian Branch, Royal Asiatic Society 26(2).

Vayda, Andrew P.
1979 Human Ecology and Economic Development in Kalimantan and Sumatra. Borneo Research Bulletin 11(1):23-32.

Wheatley, P.
1961 The Golden Khersonese: Studies in the Historical Geography of the Malayan Peninsula Before A.D. 1500. Kuala Lumpur.

Wikkramatileke, Rudolph
1958 Mukim Pulau Rusa: Land Use in a Malayan Riverine Settlement. The Journal of Tropical Geography 7:1-31.

Winstedt, R. O.
1927 The Great Flood, 1926. Journal of the Malaysian Branch, Royal Asiatic Society 5:295-303.

Index